T0258422

Advanced Research in Polyurethane

Advanced Research in Polyurethane

Edited by **Linda Cartman**

New York

Published by NY Research Press,
23 West, 55th Street, Suite 816,
New York, NY 10019, USA
www.nyresearchpress.com

Advanced Research in Polyurethane
Edited by Linda Cartman

International Standard Book Number: 978-1-63238-019-7 (Hardback)

Printed in the United States of America.

Contents

Preface

In my initial years as a student, I used to run to the library at every possible instance to grab a book and learn something new. Books were my primary source of knowledge and I would not have come such a long way without all that I learnt from them. Thus, when I was approached to edit this book; I became understandably nostalgic. It was an absolute honor to be considered worthy of guiding the current generation as well as those to come. I put all my knowledge and hard work into making this book most beneficial for its readers.

This book is a compilation of interesting work on Polyurethane. It talks about the properties, characterization, and structures of polyurethane as well as information regarding Bio-based PU, including the R&D in these areas. The topics introduce the readers to the known as well as provides unknown applications of PU, like PU for fuel binding, alkanolamide PU coatings and foams, extraction of metals, and others. It aims to serve a wider audience including readers from the fields of industrial chemistry, polymer chemistry and materials chemistry.

I wish to thank my publisher for supporting me at every step. I would also like to thank all the authors who have contributed their researches in this book. I hope this book will be a valuable contribution to the progress of the field.

Editor

Introduction

Polyurethane: An Introduction

Eram Sharmin and Fahmina Zafar

Additional information is available at the end of the chapter

1. Introduction

1.1. History of polyurethane

The discovery of polyurethane [PU] dates back to the year 1937 by Otto Bayer and his coworkers at the laboratories of I.G. Farben in Leverkusen, Germany. The initial works focussed on PU products obtained from aliphatic diisocyanate and diamine forming polyurea, till the interesting properties of PU obtained from an aliphatic diisocyanate and glycol, were realized. Polyisocyanates became commercially available in the year 1952, soon after the commercial scale production of PU was witnessed (after World War II) from toluene diisocyanate (TDI) and polyester polyols. In the years that followed (1952-1954), different polyester-polyisocyanate systems were developed by Bayer.

Polyester polyols were gradually replaced by polyether polyols owing to their several advantages such as low cost, ease of handling, and improved hydrolytic stability over the former. Poly(tetramethylene ether) glycol (PTMG), was introduced by DuPont in 1956 by polymerizing tetrahydrofuran, as the first commercially available polyether polyol. Later, in 1957, BASF and Dow Chemical produced polyalkylene glycols. Based on PTMG and 4,4'-diphenylmethane diisocyanate (MDI), and ethylene diamine, a Spandex fibre called Lycra was produced by Dupont. With the decades, PU graduated from flexible PU foams (1960) to rigid PU foams (polyisocyanurate foams-1967) as several blowing agents, polyether polyols, and polymeric isocyanate such as poly methylene diphenyl diisocyanate (PMDI) became available. These PMDI based PU foams showed good thermal resistance and flame retardance.

In 1969, PU Reaction Injection Moulding [PU RIM] technology was introduced which further advanced into Reinforced Reaction Injection Moulding [RRIM] producing high performance PU material that in 1983 yielded the first plastic-body automobile in the United States. In 1990s, due to the rising awareness towards the hazards of using chloroalkanes as blowing agents (Montreal protocol, 1987), several other blowing agents outpoured in the market (e.g., carbon dioxide, pentane, 1,1,1,2-tetrafluoroethane, 1,1,1,3,3-pentafluoropropane). At the same time, two-pack PU, PU- polyurea spray coating technology came into foreplay, which bore significant advantages of being moisture insensitive with fast reactivity. Then blossomed the strategy of the utilization of vegetable oil based polyols for the development of PU. Today, the world of PU has come a long way from PU hybrids, PU composites, non-isocyanate PU, with versatile applications in several diverse fields. Interests in PU arose due to their simple synthesis and application protocol, simple (few) basic reactants and superior properties of the final product. The proceeding sections provide a brief description of raw materials required in PU synthesis as well as the general chemistry involved in the production of PU.

2. Raw materials

PU are formed by chemical reaction between a di/poly isocyanate and a diol or polyol, forming repeating urethane groups, generally, in presence of a chain extender, catalyst, and/or other additives. Often, ester, ether, urea and aromatic rings are also present along with urethane linkages in PU backbone.

2.1. Isocyanates

Isocyanates are essential components required for PU synthesis. These are di-or polyfunctional isocyanates containing two or more than two –NCO groups per molecule. These can be aliphatic, cycloaliphatic, polycyclic or aromatic in nature such as TDI, MDI, xylene diisocyanate (XDI), meta-tetramethylxylylene diisocyanate (TMXDI), hydrogenated xylene diisocyanate (HXDI), naphthalene 1,5-diisocyanate (NDI), p-phenylene diisocyanate (PPDI), 3,3'-dimethyldiphenyl-4, 4'-diisocyanate (DDDI), 1,6 hexamethylene diisocyanate (HDI), 2,2,4-trimethylhexamethylene diisocyanate (TMDI), isophorone diisocyanate (IPDI), 4,4'-dicyclohexylmethane diisocyanate ($H_{12}MDI$), norbornane diisocyanate (NDI), 4,4'-dibenzyl diisocyanate (DBDI). Figure 1 shows examples of some common isocyanates.

The isocyanate group bears cumulated double bond sequence as R-N=C=O, wherein the reactivity of isocyanate is governed by the positive character of the carbon atom (Scheme 1), which is susceptible to attack by nucleophiles, and oxygen and nitrogen by electrophiles.

If R is an aromatic group, the negative charge gets delocalized into R (Scheme 2), thus, the aromatic isocyanates are more reactive than aliphatic or cycloaliphatic isocyanates. In case of aromatic isocyanates, the nature of the substituent also determines the reactivity, i.e., electron attracting substituents in ortho or para position increase the reactivity and electron donating substituents lower the reactivity of isocyanate group. In diisocyanates, the presence of the electron attracting second isocyanate increases the reactivity of the first isocyanate; para substituted aromatic diisocyanates are more reactive that their ortho analogs primarily attributed to the steric hindrance conferred by the second –NCO functionality. The reactivities of the two-NCO groups in isocyanates also differ with respect to each other, based on the position of –NCO groups. For example, the two-NCO groups in IPDI differ in their reactivity due to the difference in the point of location of –NCO groups. TMXDI serves as an aliphatic isocyanate since the two isocyanate groups are not in conjugation with the aromatic ring. Another isocyanate of increasing interests is vinyl terminated isocyanate since along with the –NCO group, the extra vinyl group provides sites for crosslinking (Figure 2).

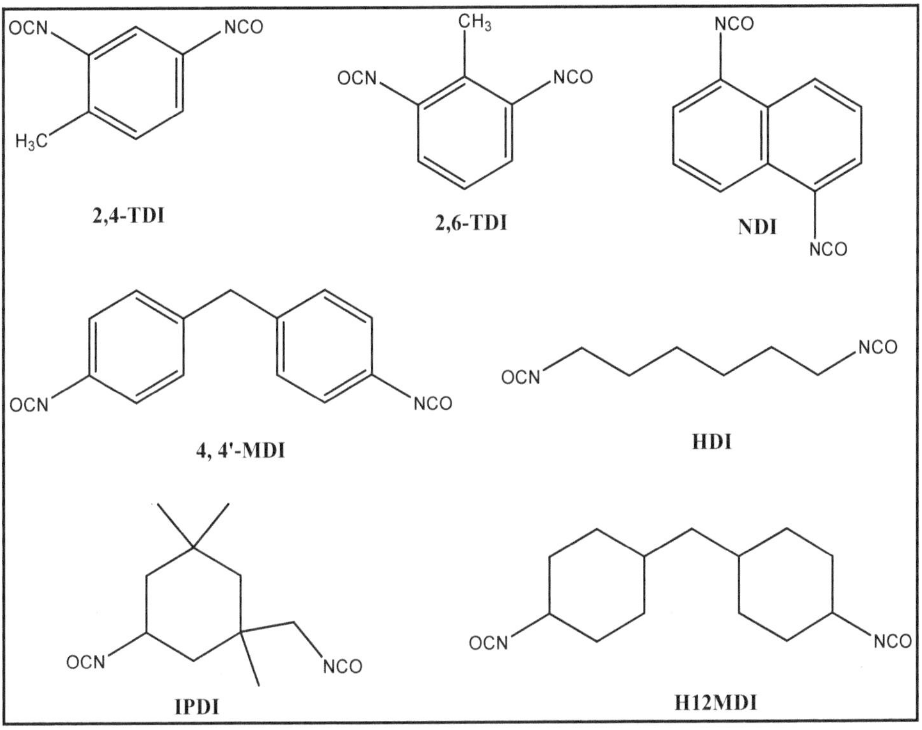

Figure 1. Common isocyanates

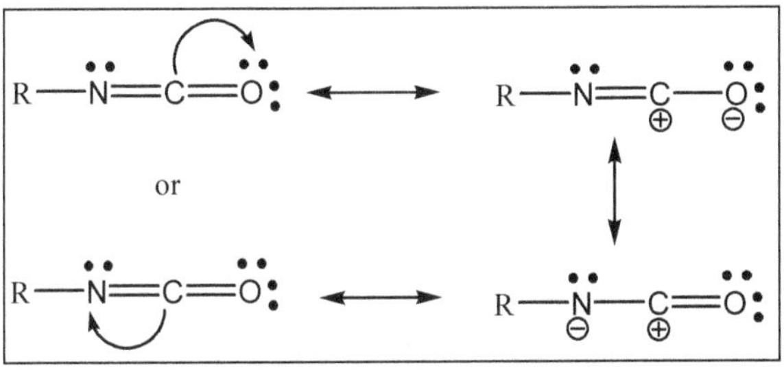

Scheme 1. Resonance in isocyanate

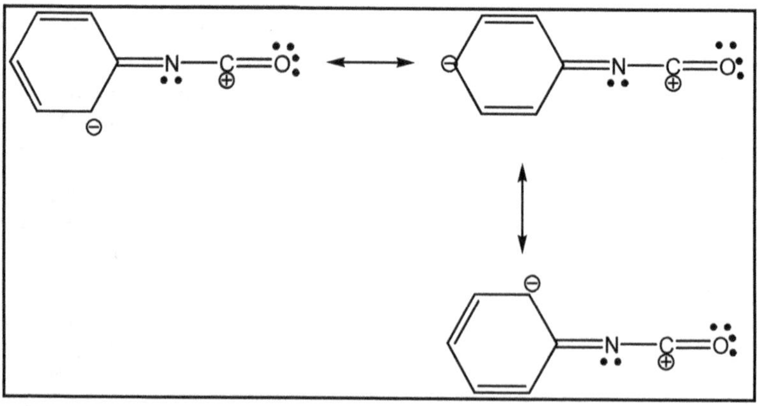

Scheme 2. Resonance in aromatic isocyanate

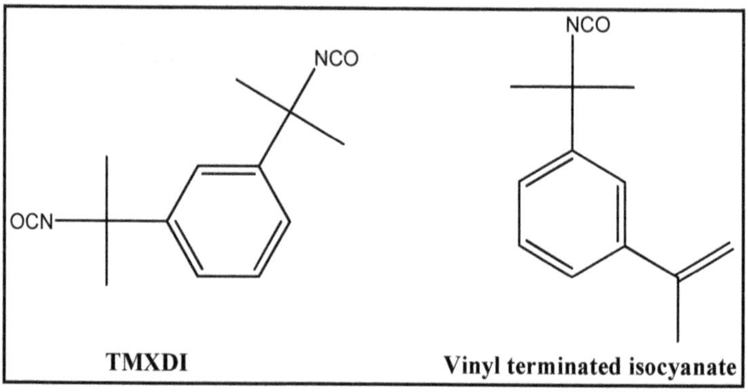

Figure 2. Other isocyanates

Polyisocyanates such as triisocyanates derived as TDI, HDI, IPDI adducts with trimethylolpropane (TMP), dimerized isocyanates termed as uretdiones, polymeric MDI, blocked isocyanates (where alcohols, phenols, oximes, lactams, hydroxylamines are blocking agents) are also used in PU production. Lately, fatty acid derived isocyanates are also prepared via Curtius rearrangement with view to produce entirely biobased PU. The choice of the isocyanate for PU production is governed by the properties required for end-use applications. To prepare rigid PU, aromatic isocyanates are chosen, however, PU derived from these isocyanates show lower oxidative and ultraviolet stabilities.

2.2. Polyols

Substances bearing plurality of hydroxyl groups are termed as spolyols. They may also contain ester, ether, amide, acrylic, metal, metalloid and other functionalities, along with hydroxyl groups. Polyester polyols (PEP) consist of ester and hydroxylic groups in one backbone. They are generally prepared by the condensation reaction between glycols, i.e., ethylene glycol, 1,4-butane diol, 1,6-hexane diol and a dicarboxylic acid/anhydride (aliphatic or aromatic). The properties of PU also depend upon the degree of cross-linking as well as molecular weight of the starting PEP. While highly branched PEP result in rigid PU with good heat and chemical resistance, less branched PEP give PU with good flexibility (at low temperature) and low chemical resistance. Similarly, low molecular weight polyols produce rigid PU while high molecular weight long chain polyols yield flexible PU. An excellent example of naturally occurring PEP is Castor oil. Other vegetable oils (VO) by chemical transformations also result in PEP. PEP are susceptible to hydrolysis due to the presence of ester groups, and this also leads to the deterioration of their mechanical properties. This problem can be overcome by the addition of little amount of carbodiimides. Polyether polyols (PETP) are less expensive than PEP. They are produced by addition reaction of ethylene or propylene oxide with alcohol or amine starters or initiators in presence of an acid or base catalyst. PU developed from PETP show high moisture permeability and low Tg, which limits their extensive use in coatings and paints. Another example of polyols is acrylated polyol (ACP) made by free radical polymerization of hydroxyl ethyl acrylate/methacrylate with other acrylics. ACP produce PU with improved thermal stability and also impart typical characteristics of acrylics to resultant PU. These PU find applications as coating materials. Polyols are further modified with metal salts (e.g., metal acetates, carboxylates, chlorides) forming metal containing polyols or hybrid polyols (MHP). PU obtained from MHP show good thermal stability, gloss and anti-microbial behavior. Literature reports several examples of VO based PEP, PETP, ACP, MHP used as PU coating materials. Another example is VO derived fatty amide diols and polyols (described in detail in chapter 20 Seed oil based polyurethanes: an insight), which have served as excellent starting materials for the development of PU. These PU have shown good thermal stability and hydrolytic resistance due to the presence of amide group in the diol or polyol backbone.

2.3. Additives

Along with a polyol and an isocyanate, some additives may also be required during PU production, primarily to control the reaction, modify the reaction conditions, and also to finish or modify the final product. These include catalysts, chain extenders, crosslinkers, fillers, moisture scavengers, colourants and others. In PU production, catalysts are added to promote the reaction to occur at enhanced reaction rates, at lower temperatures, for deblocking the blocked isocyanates, for decreasing the deblocking and curing temperatures and times. A number of aliphatic and aromatic amines (e.g., diaminobicyclooctane-DABCO), organometallic compounds (e.g., dibutyltin dilaurate, dibutyltin diacetate), alkali metal salts of carboxylic acids and phenols (calcium, magnesium, strontium, barium, salts of hexanoic, octanoic, naphthenic, linolenic acid) are used as catalysts. In case of tertiary amines, their catalytic activity is determined by their structure as well as their basicity; catalytic activity increases with increased basicity and decreases with the steric hindrance on the nitrogen atom of amine. They promote their catalytic action by complex formation between amine and isocyanate, by donating the electrons on nitrogen atom of tertiary amine to the positively charged carbon atom of the isocyanate. Metal catalysts bear superiority over tertiary amines because they are comparatively less volatile and less toxic. Metals catalyse the isocyanate-hydroxyl reaction by complex formation with both isocyanate and hydroxyl groups. The positive metal centre interacts with electron rich oxygen atom of both the isocyanate and hydroxyl groups forming an intermediate complex, which by further rearrangement results in the formation of urethane bonds. Difunctional low molecual weight diols (ethylene glycol, 1,4-butanediol, 1,6-hexanediol), cyclohexane dimethanol, diamines, hydroxyl-amines (diethanolamine and triethanolamine) are used as chain extenders in PU synthesis while those with functionality 3 or > 3 are used as crosslinkers. Since isocyanates are too sensitive to moisture or water even in traces, moisture scavengers, which react more readily with water than an isocyanate, are incorporated to cut off/eliminate the involvement of water during PU synthesis, e.g., oxazolidine derivatives, zeolite type molecular sieves. Blowing agents are used to produce PU foams with cellular structures by foaming process (e.g., hydrocarbons, CO_2, hydrazine).

3. Chemistry of PU

PU are carbonic acid derivatives. The older term for them is an ester of a substituted carbamic acid, polycarbamate, from carbamic acid. PU are formed by (i) the condensation polymerization reaction of bischloroformates with diamine (Scheme 3) and (ii) addition polymerization reaction of diisocyanates with di or polyfunctional hydroxy compounds, or other compounds having a plurality of active hydrogen atom (Scheme 4). The latter method is more important from the industrial point of view since in this method no by-product is formed.

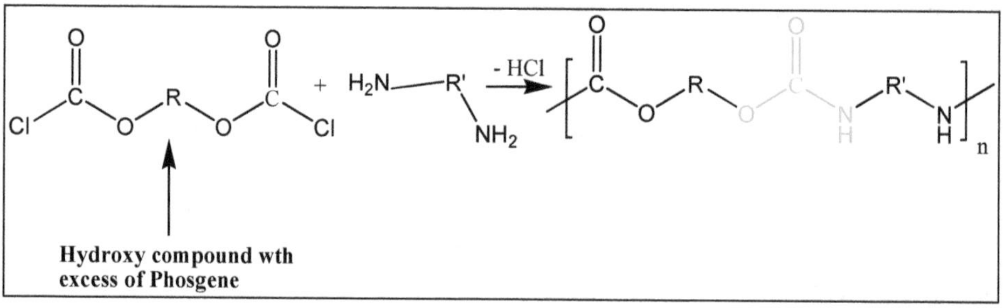

Hydroxy compound wth excess of Phosgene

Scheme 3. Reaction of bischloroformate with diamine

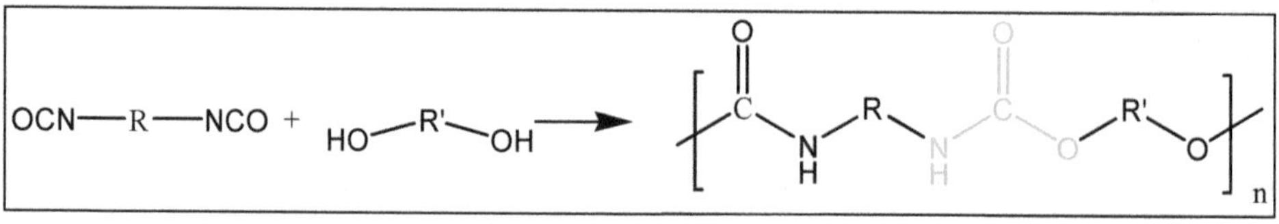

Scheme 4. Reaction of diisocyanate with di or poly hydroxy compound

The isocyanate reaction offers the possibility of producing tailor-made polymeric product ranging from fibres to rubber. Generally, the isocyanate reactions are divided into two classes, (a) addition (primary and secondary) reaction with compound containing active hydrogen (Schemes 5 and 6), (b) self-addition reaction (Scheme 7). In some of the reactions, CO_2 is released which assists in the formation of PU foams.

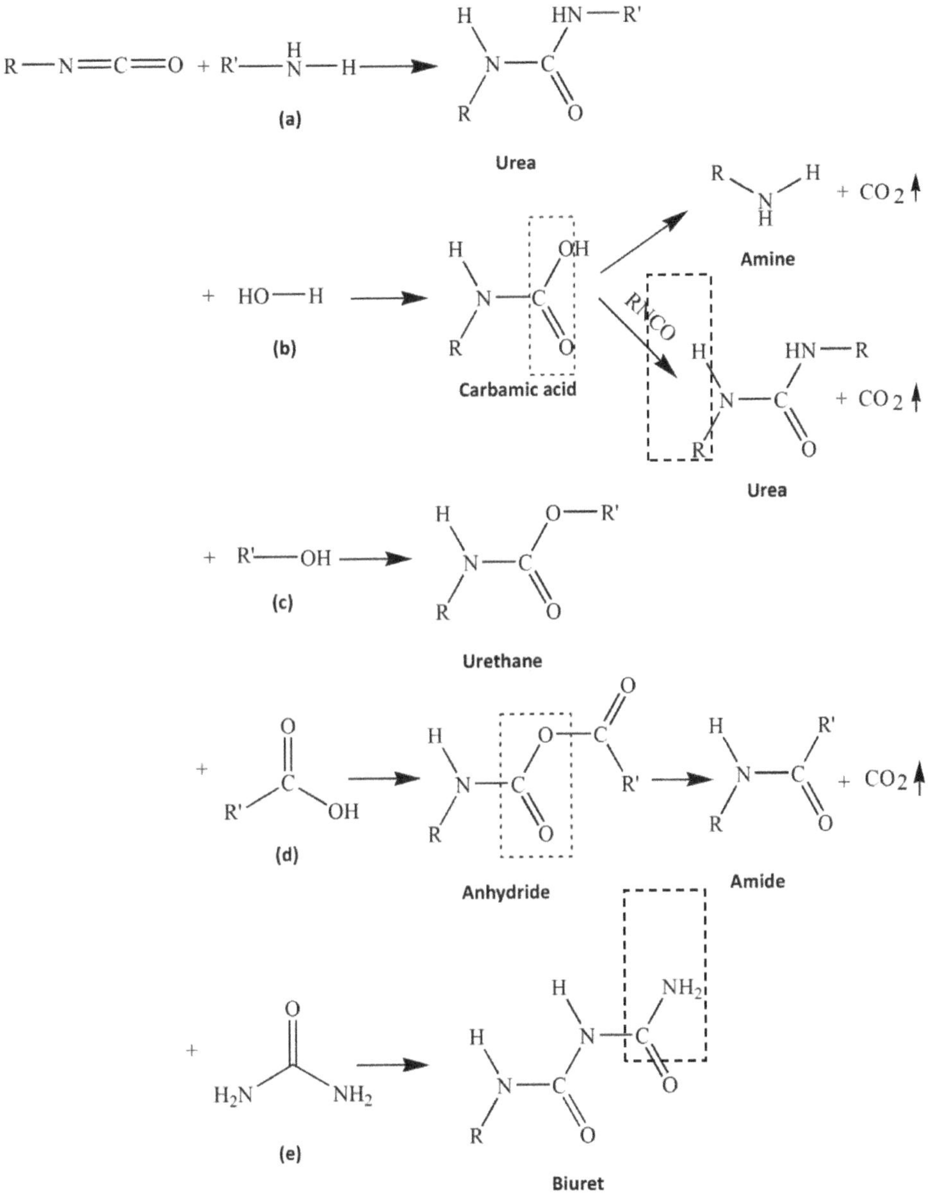

Scheme 5. Primary addition reactions of isocyanate with (a) amine, (b) water, (c) alcohol, (d) carboxylic acid, (e) urea.

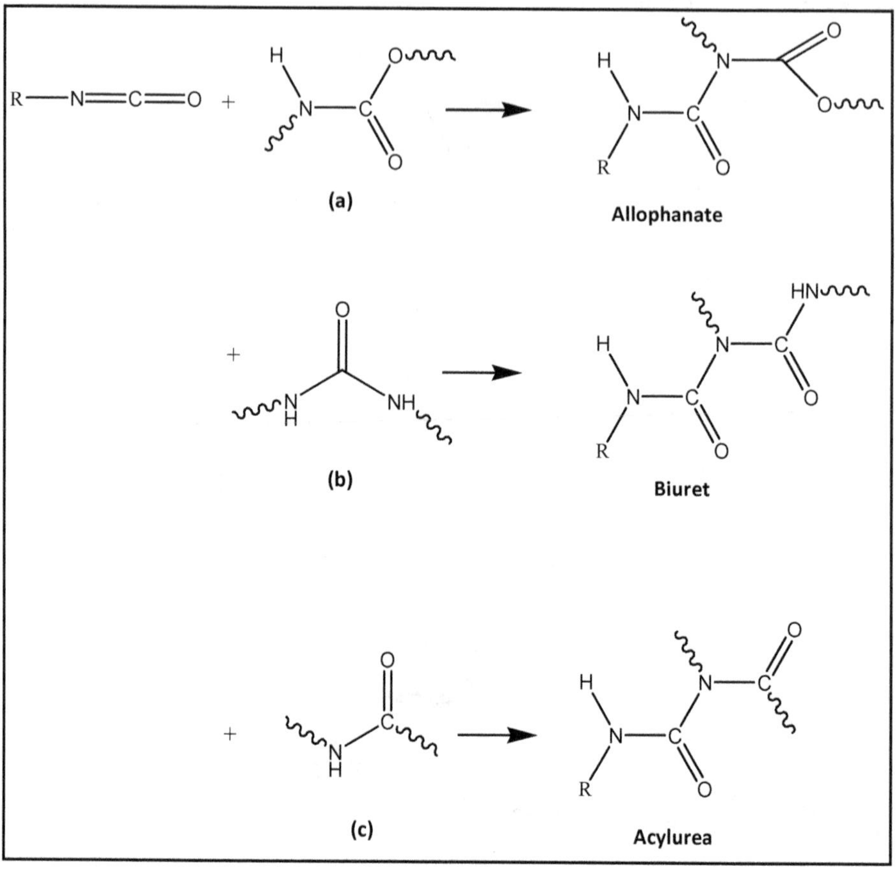

Scheme 6. Secondary addition reactions of isocyanate with (a) polyurethane, (b) polyurea and (c) polyamide

Wurts in 1848 discovered the basic reaction of isocyanate (Scheme 4). He found that isocyanates having the structure R-N=C=O, where R= alkyl or aryl group, react rapidly at room temperature with compounds containing active hydrogen atoms, like amine, water, alcohol, carboxylic acid, urethanes and ureas (Scheme 8).

It is observed that a linear PU is formed when a diisocyanate react with diol whilst branched or cross-linked PU results with the reaction of polyhydric compound (polyol). The branched or cross- linked PU are also formed when a compound containing three or more isocyanate groups reacts with a diol; however, this approach is of limited commercial importance.

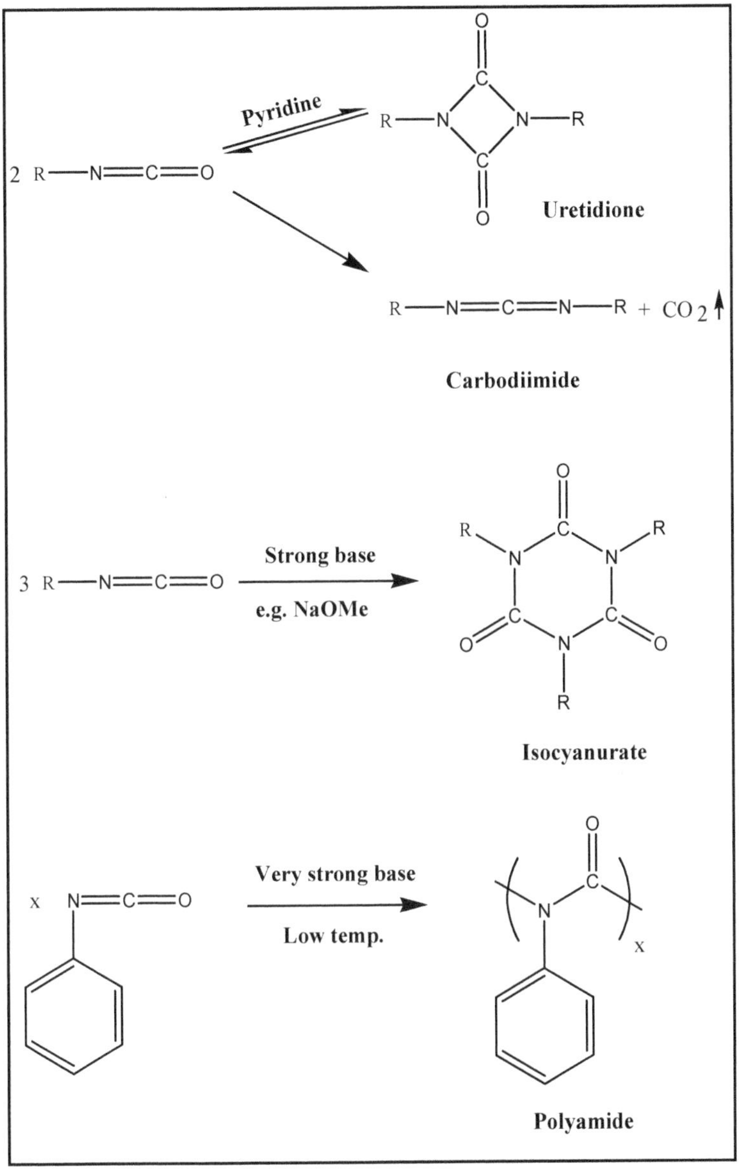

Scheme 7. Self -addition reactions of isocyanate

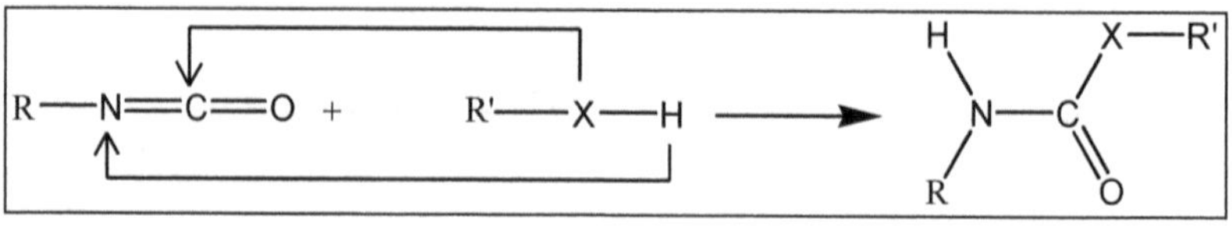

Scheme 8. Reaction of isocyanate with active hydrogen compound

4. Mechanism

The reaction of an isocyanate with active hydrogen compounds is carried out with or without a catalyst. The self-addition reactions of isocyanates do not usually proceed as readily as reactions with active hydrogen compounds.

4.1. Reaction in the absence of a catalyst

The active compound itself acts catalytically in the reaction as follows (Scheme 9).

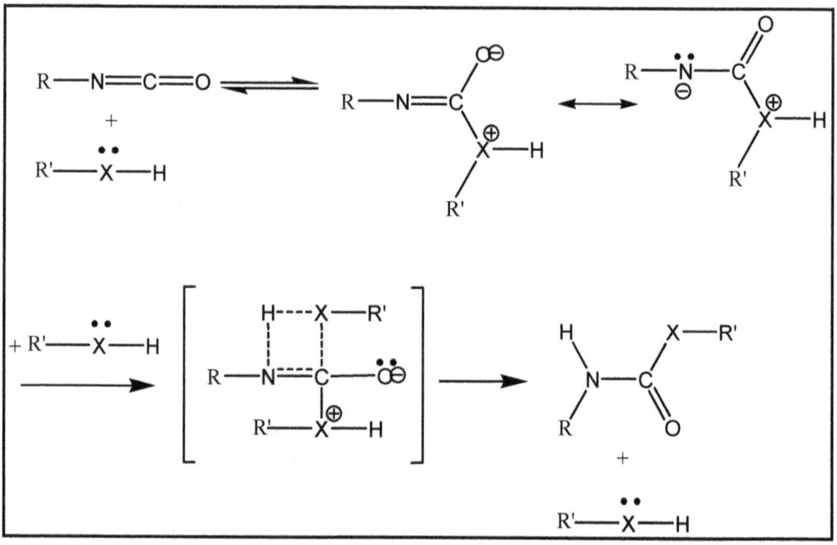

Scheme 9. Isocyanate reaction in the absence of a catalyst

As given in Scheme 9, in the reactions proceeding in the absence of a catalyst, the electrophilic carbon of the isocyanate is attacked by the nucleophilic centre of the active hydrogen compound; hydrogen is added to –NCO group. The reactivity of the –NCO groups is increased due to the presence of the electron withdrawing groups, and decreases by the electron donating groups. While the aromatic isocyanates are more reactive than the aliphatic isocyanates, steric hindrance at –NCO or HXR' groups reduce the reactivity.

The order of reactivity of active hydrogen compounds with isocyanates in uncatalyzed systems is as follows:

Aliphatic amines> aromatic amines> primary alcohols> water>secondary alcohol> tertiary alcohol> phenol> carboxylic acid> ureas> amides>urethanes.

4.2. Reaction in the presence of a catalyst

The isocyanate reactions of class (a) are also extremely susceptible to catalysis. The various isocyanate reactions are influenced to different extents by different catalysts. Many commercial applications of isocyanates utilize catalysed reactions. Tertiary amines, metal compounds like tin compounds (as mentioned earlier in the chapter) are most widely used catalysts for the reaction (Schemes 10 and 11). The mechanisms are similar to that of the uncatalyzed reaction (Scheme 9).

The tertiary amines and metal salts catalyse the reaction as follows:

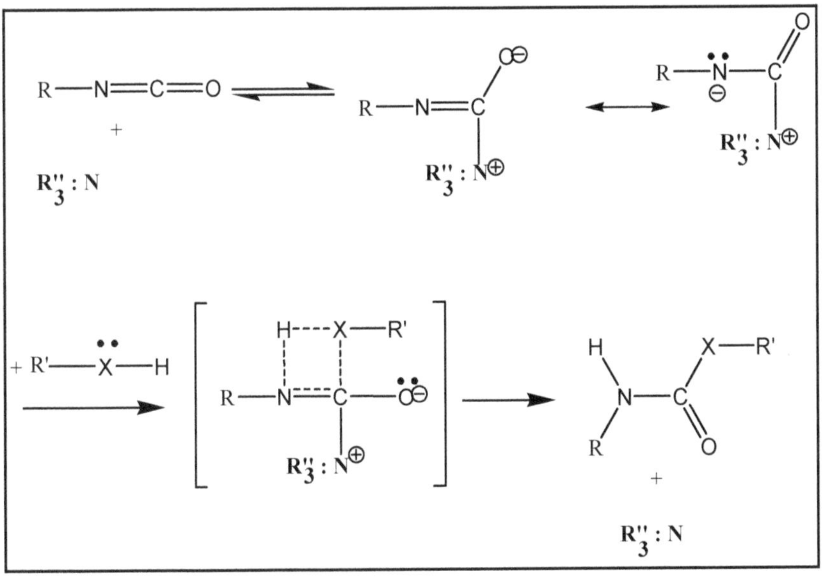

Scheme 10. Tertiary amine catalysed reaction

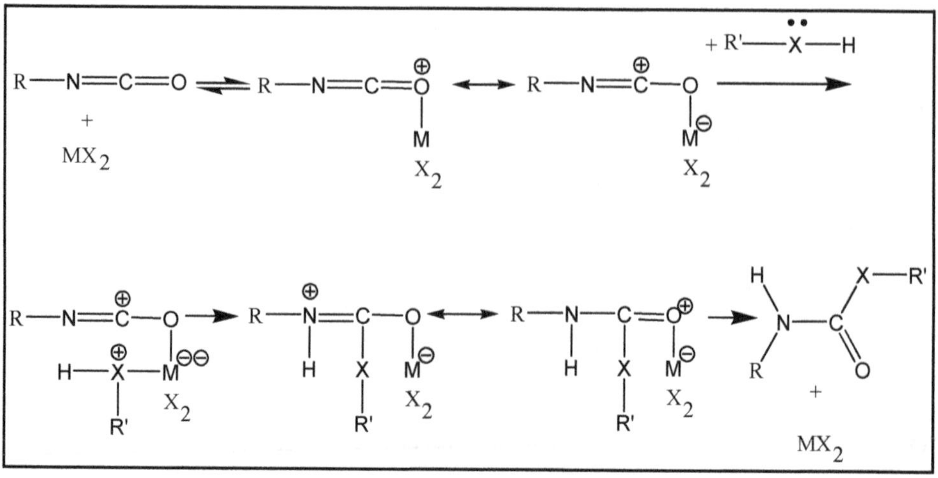

Scheme 11. Metal salts catalysed reaction

The catalytic activity of amines closely parallels to the base strength of the amines except when steric hindrance becomes pronounced. This catalyst is also effective for self-addition reactions while metal salt compounds generally have less influence; tin compounds are particularly poor catalysts in these reactions.

5. Hazards

Although PU are chemically inert in their fully reacted form , the risks of asthmatic symptoms arise on human exposure even in smaller concentrations due to the volatility associated with isocyanates arise the risk of asthmatic symptoms on 12 human exposure, even in smaller concentrations. On exposure to flames, hazards of ignition are feared. Isocyanates may also be sensitive on our skin. Some isocyanates may also be anticipated as carcinogens. Thus, persons working with isocyanates must be equipped with proper protection devices such as gloves, masks, respirators, goggles, and others, as precautionary measures.

6. Conclusion

PU are thermoplastic and thermoset in nature. The type, position, and structure of both the isocyanate and polyol determine the progress of PU forming reactions as well as their properties and end-use applications. Hydrogen bonding also plays a key role in determining the properties of final PU product. Due to the associated health hazards, complete precautions are necessary while working with isocyanates. PU are available as one-pack or two-pack PU. PU dispersions, waterborne PU, PU Interpenetrating

Networks PU, hybrids and composites are used in various applications such as paints and coatings, adhesives, sealants, foams, absorbents, flame retardants, fuel binders, in automobiles, in biomedical applications (urological stenting practices, carriers of antituberculosis drugs, orthopaedics), extraction of metals, grouting technologies, crashworthiness, treatment of industry wastewater, cast elastomers, and others as also discussed in proceeding chapters.

Author details

Eram Sharmin and Fahmina Zafar*
Materials Research Laboratory, Department of Chemistry,
Jamia Millia Islamia (A Central University), New Delhi, India

Acknowledgement

Dr Fahmina Zafar (Pool Officer) and Dr Eram Sharmin (Pool Officer) acknowledge Council of Scientific and Industrial Research, New Delhi, India for Senior Research Associateships against grant nos. 13(8385-A)/2010-POOL and 13(8464-A)/2011-10 POOL, respectively. They are also thankful to the Head, Department of Chemistry, Jamia Millia Islamia (A Central University), for providing support to carry out the work.

7. References

Chattopadhyay D.K., Raju K.V.S.N. Structural Engineering of Polyurethane Coatings for High Performance Applications. Progress in Polymer Science 2007; 32: 352–418.

Desroches M., Escouvois M., Auvergne R., Caillol S., Boutevin B. From Vegetable Oils to Polyurethanes: Synthetic Routes to Polyols and Main Industrial Products. Polymer Reviews 2012; 52 (1): 38–79.

Lligadas G., Ronda J.C., Galia`M., Cadiz V. Plant Oils as Platform Chemicals for Polyurethane Synthesis:Current State–of–the–Art. Biomacromolecules 2010; 11: 2825–2835.

Malcolm P S. Polymer Chemistry An Introduction. 3rd Edn. New York: Oxford University Press, Oxford; 1999.

Nylen P., Sunderland E. Modern Surface Coatings. London: John Wiley & Sons; 1965.

Paul, S. Surface Coating–Science and Technology. New York: John Wiley & Sons; 1985.

Petrović Z. S. Polyurethanes from Vegetable Oils. Polymer Reviews 2008; 48:109–155.

---

* Corresponding Author

Pfister D.P., Xia Y., Larock R.C. Recent Advances in Vegetable Oil–based Polyurethanes. Chem Sus Chem 2011; 4(6):703–17.

Saunders K J. Organic Polymer Chemistry. 2nd Edn. New York: Chapman & Hall; 1981.

http://en.wikipedia.org/wiki/Polyurethane (accessed on 11th July 2012)

Synthesis and Properties

The Modification of Polyurethanes by Highly Ordered Coordination Compounds of Transition Metals

Ruslan Davletbaev, Ilsiya Davletbaeva and Olesya Gumerova

Additional information is available at the end of the chapter

1. Introduction

One of the ways to influence the chemical structure of polyurethanes is to use metal complex systems based on transition metal chlorides for their synthesis. The significance of this trend is conditioned by the ability of metal complexes to order the macromolecular chains, as well as affect the electrical properties of polyurethanes (Davletbaeva et al., 1996, 2001).

The synthesis of metal coordination polymers is a way of affecting the processes of crosslinking of macrochains; interchain and intraionic interactions; and, thereby, preparing polymer materials with special properties (Dirk et al., 1986; Kingsborough & Swager, 1999; Reynolds et al., 1985; Thuchide & Nishide, 1977; Wang & Reynolds, 1988). From the standpoint of designing materials with electric and magnetic properties, it is promising to form in a polymer matrix chains of transition metal ions bound by exchange interaction.

Conventional methods for the creation of interactions of this type in a polymer are primarily based on the presence of certain units in a macromolecule, e.g., those including the phthalocyanine and azomethine moieties. For example, metal atoms in metal phthalocyanine liquid crystalline complexes are bound to one another by chloride bridges and play the role of a spacer between phthalocyanine units, thus promoting overlap of electronic orbitals of parallel molecules (Shirai et al., 1977, 1979).

As a result, the electric conductivity of metal-coordination polymers obtained on the basis of these complexes is increased by a few orders of magnitude relative to undoped systems. The coordination bonding of comb-like liquid crystalline polymers can also give rise to stacked structures. Interaction between metal ions is revealed in such polymers, which is realized owing to the association of metal ions in an indirect manner, through ester oxygen bridges.

However, this approach is seriously limited and cannot be used for the creation of stacked metal-coordinated fragments in a disordered polymer matrix (Brostow, 1990; Carrher, 1981; Serrano & Oriol, 1995).

Metal complex structuring is promising in terms of the influence on properties of polyurethanes. The significance of this trend is conditioned by the ability of metal complexes to order the macromolecular chains, as well as affect the electrical properties of polyurethanes.

2. Coordination compounds based on the aromatic isocyanates and copper (II) chloride for the synthesis of polyurethanes

In the early studies (Davletbaeva et al., 1998) devoted to imparting special properties to polyurethanes by their coordination bonding, it was shown that the interaction of aromatic isocyanates with copper chloride ($CuCl_2$) in the acetone medium in the presence of trace amounts of water proceeded as a sequence of chemical transformations including simple addition and decomposition reactions, redox processes, and subsequent complexation (Fig. 1).

The ultimate products are polynuclear complexes of azoaromatic compounds, in which copper ions occurring in two variable oxidation states are connected by chloride bridges (Fig. 1). As it is seen from the structural formulas, some copper ions are stabilized at the initial degree of oxidation due to the formation of heterovalent pairs connected by chloride bridges. Free isocyanate groups present in these compounds are able to react with oligodiols of different nature. Polyurethanes obtained in such manner form polymer network by coordination bonding of urethane groups and azogroups which are the part of a macrochain.

It is shown that the chloride bridges are replaced by heteroatoms that are present in polymer chains while metal ions are coordinatively bound to macromolecules to crosslink them and, occurring in two interacting variable oxidation states, to form local centers of exchange interactions. As a result of electron transfer from one local coordinated unit to another, which is mediated by electron-donating groups, such as an ester group, the conductivity of polyurethane increases by several orders of magnitude.

3. Reactions of aromatic isocyanates and urethane prepolymers with coordination compounds of iron (III)

Further studies established that similar structural units could be formed even in the polyurethane matrix itself as a result of its modification with metal complexes synthesized for this purpose. A characteristic feature of these coordination compounds is the presence of chloride-bridged metal ions in their structure. One of such crosslinking metal complex system was prepared by the reaction of iron chloride ($FeCl_3$) with ethanolamine (EA). It was found that reactions involving aromatic isocyanates and EA in the coordination sphere of the iron ion (III) led to the formation of azoaromatic compounds shown in figure 2.

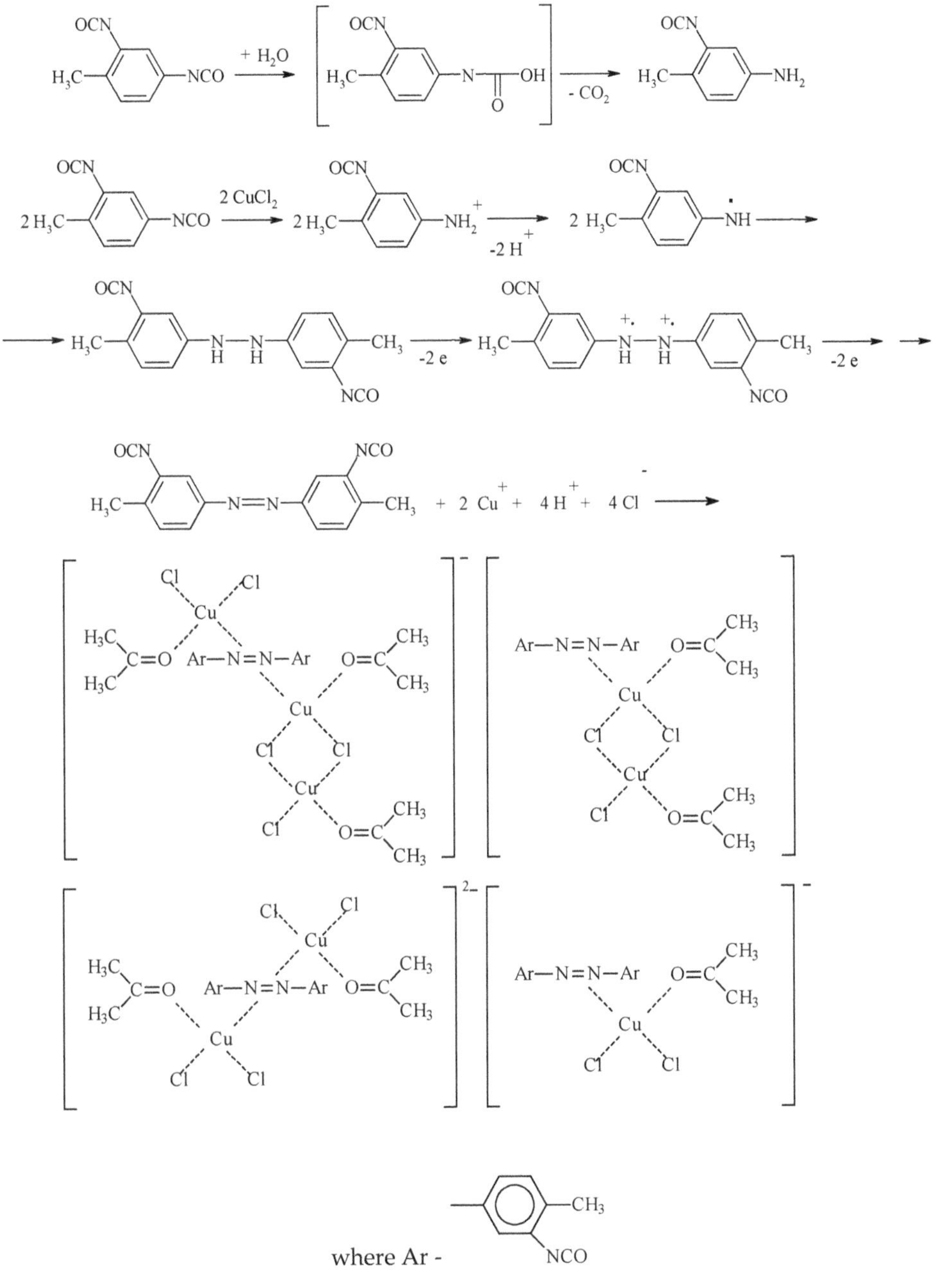

Figure 1. Formation of polynuclear complexes of azoaromatic compounds.

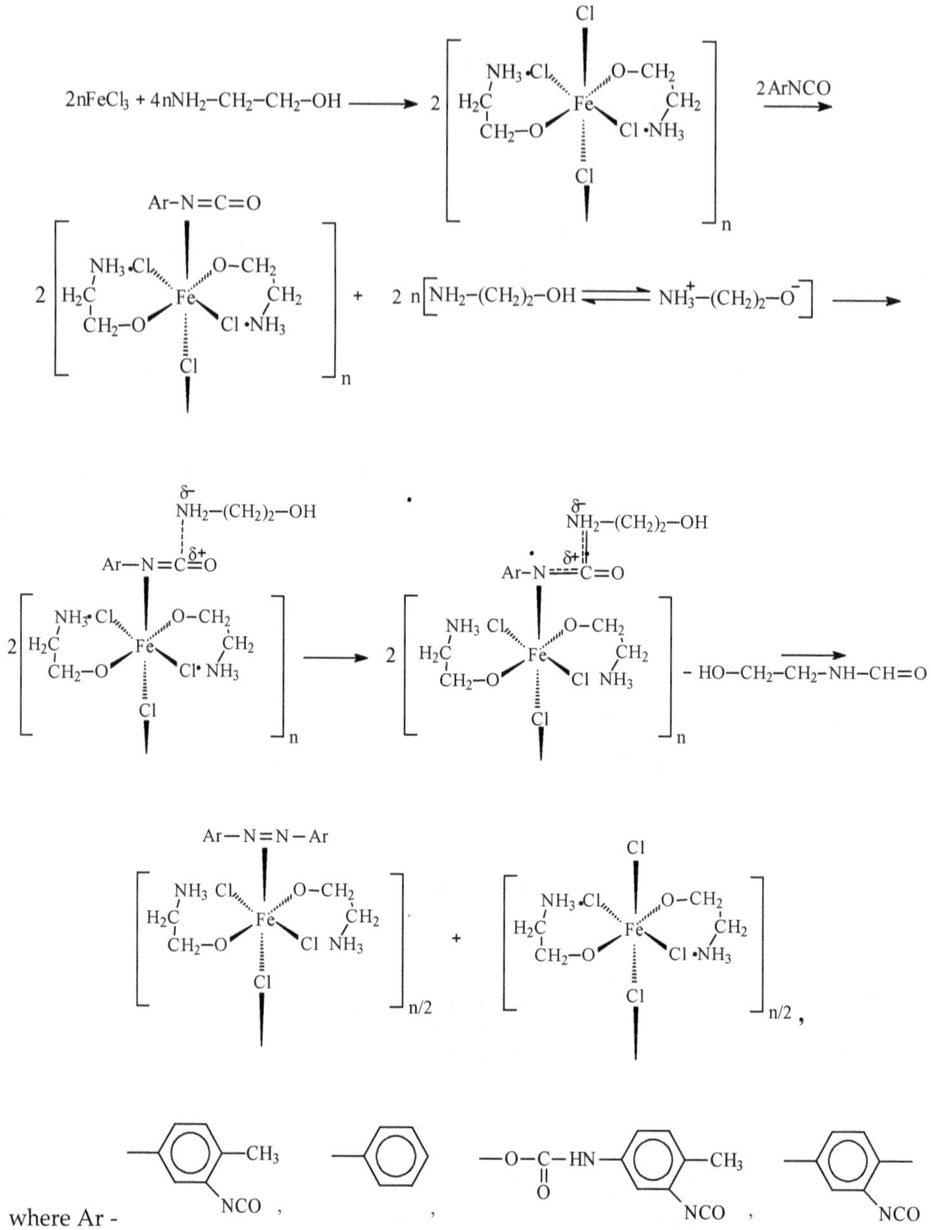

Figure 2. Scheme of formation of azoaromatic compounds.

The continuation of this reaction is the formation of stack coordination compounds in which metal ions form coordination bonds with azogroups (Fig. 3).

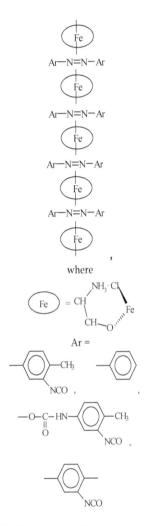

Figure 3. Formation of stack coordination compounds.

Mössbauer studies of iron coordinated compounds reveal that the magnetic ordering is observed at relatively high temperature (80 K). Mössbauer spectrum obtained in transmission geometry at the temperature of the sample being equal to 80 K consists of two components, namely, a doublet in the middle of the spectrum corresponding to the residues of the initial FeCl₃, dissolved in the matrix (less than 10% of the total area under the spectrum) and the magnetically ordered component with a hyperfine field average value of about 430 kE and the isomer shift, indicating a high-spin state of the Fe (III) ion. The absence of partial component in the middle of the spectrum is caused by the long average size of supramolecular structures chains (Fig. 4).

The Mössbauer study confirms the columnar structure of the obtained metal complexes, the possibility of their fixation in a flexible-chain polymer matrix containing electron-donating groups, and the existence of magnetic ordering at temperatures below 70 K. Sizes of very thin magnetic fields correspond to high-spin state of iron(III) ions (S=5/2).

It is known that to achieve the effect of magnetic ordering it is necessary that the chain of interacting ions of iron (III) should be long enough and combine up to a few thousand ions. The criterion for judging the length of the chain of magnetic ordered iron ions is relatively high blocking temperature of supermagnetism and the barrier value of effective anisotropy field. In our view, due to significant anisotropy in the structure of the complex the most likely assumption is the increasing of the potential barrier with the increase of chain length. The longer the chain, the higher its strength and the ability to build columns in a polymer matrix, and the specific properties of the structured polymer matrix are more pronounced as well.

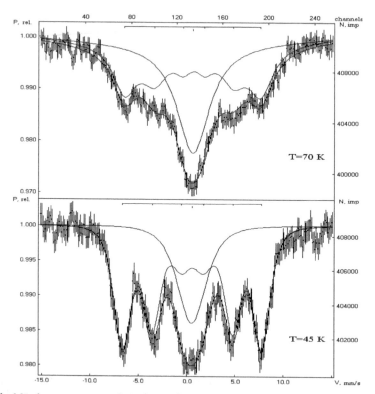

Figure 4. The Mössbauer spectrum of metal complex system based on FeCl₃ and EA.

The metal complex system is used for structuring urethane prepolymer containing terminal isocyanate groups. Considering the high flexibility of the urethane prepolymer chain, it can be assumed that the urethane groups will be coordinately bound with iron ions. The result of this interaction should be the formation of columnar structures directly in the polymer matrix. Mössbauer studies of urethane prepolymer confirm these assumptions. The resulting spectra are also of superparamagnetic nature at temperatures below 47K (Davletbaeva et al., 2006).

When polyurethanes are modified by coordination compounds synthesized on the basis of $FeCl_3$ and EA, the minimum values of specific volume electrical resistance (about 10^8 Ohm•sm) are recorded in the concentration area of 0,1% in terms of iron chloride. It should be noted that the ions of iron (III) in the above reactions do not change the oxidation level.

4. Modification of urethane prepolymer by coordination compounds of copper (I, II)

In one of the worked out highly ordered coordination compounds of transition metals for modification of polyurethanes N,N'-Diethylhydroxylamine (DEHA) is used as a ligand exhibiting the properties of a reducing agent. The most appropriate transition metal compound is copper (II) chloride. Some of the Cu (II) ions interacting with DEHA reduce the oxidation level and turn into Cu (I) (see Fig. 5).

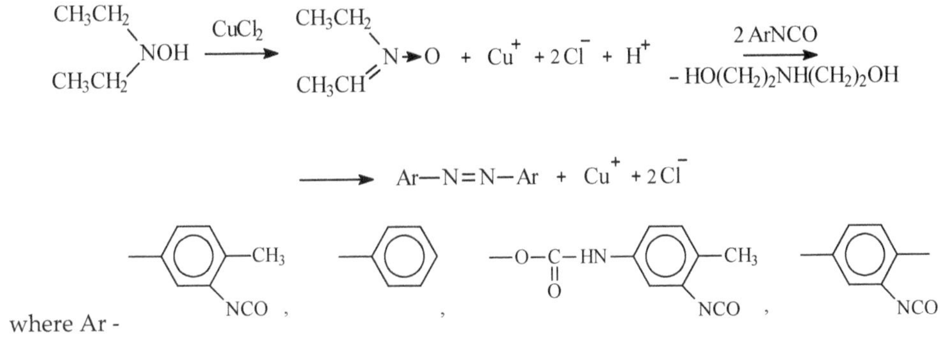

Figure 5. The mechanism of interaction of copper (II) chloride with DEHA.

It was ascertained that the metal complex system showed the ability to interact with aromatic isocyanates to form azoaromatic compounds. The result was the formation of columnar coordination compounds (Fig. 6).

It was established, that metal-complex modification of polyurethanes results in the change of their physicomechanical properties and spasmodic reduction (by 3-4 orders) of volumetric and superficial electric resistance. The reduction of the specific bulk electrical resistance by 3-4 orders is the most significant effect accompanying the metal-complex binding of polyurethanes. In this case the electrical resistance falls spasmodically depending on the nature of flexible chains of polyurethanes, the range of transition -metal ions concentrations. If the content of metal-complex modifying agent is increased further, the electrical resistance increases to some extent. The main role in the mechanism of charge transmission in metal-coordinated polyurethanes is assigned to electron-donating groups which are included in the structure of flexible chains of polyurethane matrix and the presence of transition metal ions having two degrees of oxidation in it.

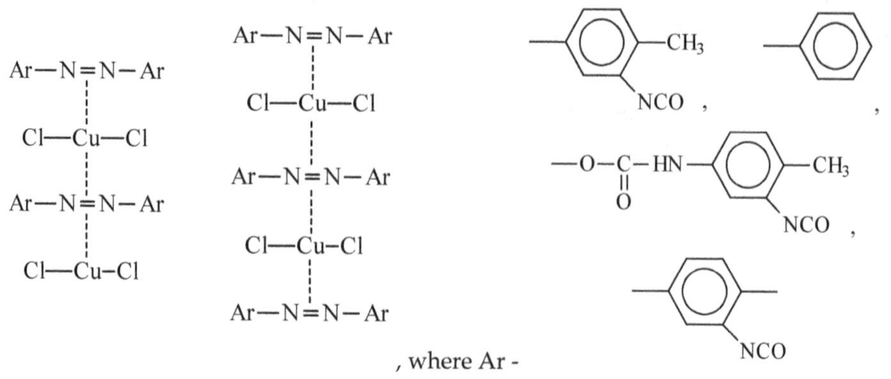

Figure 6. Formation of columnar coordination compounds.

5. Modification of polyurethanes by heteronuclear complexes based on molybdenum (V) and copper (II) chlorides

To study the effect of the coordination compounds structure on the electrophysical properties of the modified polyurethanes the heteronuclear metal complexes based on transition metals of IV and V periods were synthesized. CuCl$_2$ was used as the chloride of 3d-metal, and MoCl$_5$ was used as the chloride of the 4d-metal. DEHA was used as a ligand. Cu (II) has 3d-orbitals that are involved in coordinating binding, in the case of Mo (V) this role is performed by 4d-orbitals. Therefore, it is assumed that when heteronuclear complexes of the columnar structure are formed, where the CuCl$_2$ is in excess, the ions of molybdenum could cause the interruption of chains of exchange interactions between 3d-ions (Fig. 7). It is found that the metal complex system, obtained on the basis of [CuCl$_2$]:[DEHA] = 1:0,7 has ρ_v = 2600 Ohm·sm at room temperature, while the system based on [MoCl$_5$]:[DEHA] = 1:0,7 has ρ_v = 1300 Ohm·sm. The heteronuclear complex obtained at the ratio of [CuCl$_2$]:[MoCl$_5$]:[DEHA] = 0,9:0,1:0,7 at room temperature exhibits ρ_v = 5250 Ohm·sm, while the complex obtained at ratio of [CuCl$_2$]:[MoCl$_5$]:[DEHA] = 0,8:0,2:0,7 has already ρ_v = 73800 Ohm·sm.

Homonuclear and heteronuclear metal complexes were used for polyurethane modification. It was found that both modifying systems were able to react with the urethane prepolymer. When the metal complex structuring of polyurethanes by heteronuclear complexes based on 3d-and 4d-ions took place ϱv values remained similar to the unmodified sample. With the increase of the concentration of heteronuclear complex the specific volume electrical resistance of polyurethanes actually increased slightly (Fig. 8). Another situation is observed in the case of polyurethanes modification by homonuclear coordination compounds. In this case the introduction of metal complexes based on copper led to the decrease of ϱv by three orders. However, when polyurethanes were modified by homonuclear metal complex compounds based on Mo (V) ϱv decreased by 5 orders.

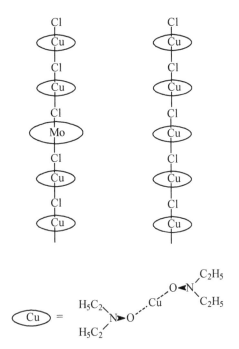

Figure 7. Scheme of the formation of columnar homonuclear and heteronuclear coordination compounds of transition metal.

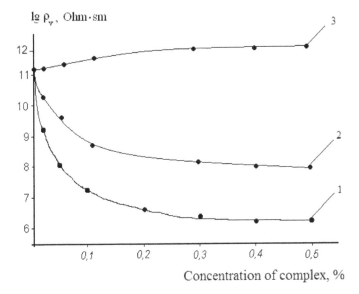

Concentration of complex, %

Figure 8. Dependence of ρ_v (Ohm·sm) of polyurethanes on the concentration (%) of metal complexes based on: 1-MoCl$_5$-DEHA; 2-CuCl$_2$-DEHA; 3-CuCl$_2$-MoCl$_5$-DEHA.

6. Catalytic properties of coordination compounds of copper in the reaction with isocyanate and urethane groups

The next step was to change the ligand composition of metal complex modifying system based on CuCl2 and DEHA. The aminopropyltriethoxysilane (AGM) was used as a part of modifying system. The use of the AGM as an additional component to the DEHA was caused by some reasons. The first reason is that the AGM is able to take part in the reactions of sol-gel synthesis, resulting in the hydrolysis of ethoxy-component and subsequent polycondensation of the forming silanol groups. The second reason is the presence of electron-donor amine groups in the AGM which are able to form complexes. Besides, amines can lead to reduction of copper (II) to copper (I). Thereby this substance is interesting in terms of the influence on the reactivity of isocyanate groups and the supramolecular structure of polyurethane, which has domain nature.

Titrimetric determination of concentration of isocyanate groups in UPTI during its interaction with metal complex system based on CuCl2, DEHA and AGM at 100ºC, was carried out. UPTI is industrial prepolymer synthesized on the base of 1 mol polyoxitetramethyleneglicol and 2 mols 2,4-toluene diisocyanate.

It was found that at relatively low concentrations of metal complex (0.01 and 0.05% in terms of CuCl2) in the first ten minutes from the start of the reaction process, the concentration of isocyanate groups started to rise, and only after that it fell. When the content of metal complex was 0.1, 0.5 and 0.75% in terms of $CuCl_2$ the concentration of isocyanate groups began to fall significantly (see fig. 9).

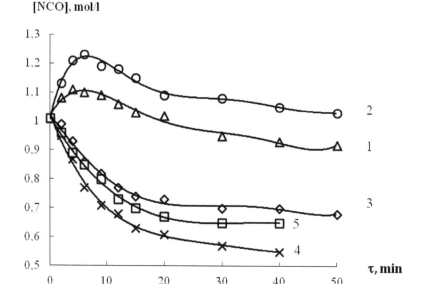

Figure 9. Isocyanate groups consumption curves in system UPTI – CuCl2-DEHA-AGM, T=100°C, at a content of CuCl2: 1 - 0,05%; 2 - 0,1%; 3 - 0,25%; 4 - 0,5%; 5 - 0,75% (wt.).

The titration data confirm the results of IR-spectroscopic studies. It is established that the absorption band at 1731 sm-1 due to the stretching vibrations of carbonyl component of urethane group decreases at low concentrations of metal complex in the first ten minutes from the start of the reaction process. At the same time the intensity of the absorption band at 2273 sm^{-1} due to the stretching vibrations of isocyanate group grows. Later the growth of the intensity of the absorption band at 1731 sm^{-1} and the decrease at 2273 sm^{-1} are observed. Besides in the first ten minutes from the start of the reaction the IR-spectroscopy shows the reduction of the intensity of the absorption band at 3293 sm^{-1} due to the stretching vibrations of N-H-bond that is a part of urethane group. At relatively high concentrations of metal complex ($\geq$0.25%) the interaction is accompanied by the growth of the intensity of the absorption band at 2120 sm-1 due to the formation of carbodiimide group.

The research suggested that at relatively low concentration of metal complex the urethane group dissociates to isocyanate and hydroxyl groups, while at high concentrations of metal complex the isocyanate groups consume to the formation of carbodiimide groups. The part of isocyanate groups is hypothetically consumed on the formation of the azoaromatic groups. It is known that it is impossible to analyze azoaromatic groups using infrared spectroscopy. In this regard, studies were carried out using electron spectroscopy.

Electronic spectrum (Fig. 10) showed absorption at 350 nm, typical for trans-azoaromatic compounds. The absorption in the area 480 nm characterizes the coordination compounds of copper (II).

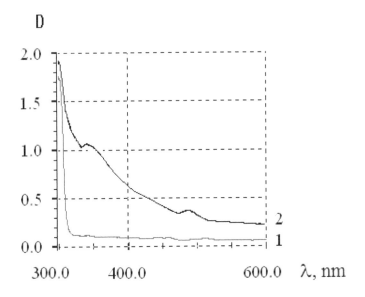

Figure 10. Electron spectrum of urethane prepolymer (1) and prepolymer (2) modified by 0.5% (wt.) metal complex system based on CuCl₂ - DEHA – AGM.

Metal complex system, derived on the basis of CuCl₂, DEHA and AGM was further used to modify polyurethanes. We measured the dependence of the volume resistivity (ρ_v) of polyurethanes on the concentration of metal complex modifier (Fig. 11).

It turned out that the use of the worked out metal complex system caused the leap of ρ_v by 4 orders (10 000 times) observed at low concentrations of metal complex 0.01%. Here we should note that while using metal complex system based on CuCl₂ and DEHA (no AGM) a stepwise drop of ρ_v was observed at much higher concentration of the complex - 0.1%.

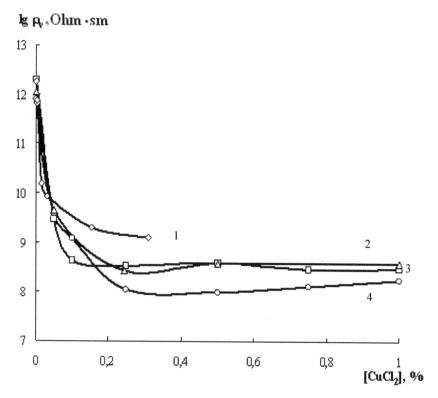

Figure 11. Volume resistivity-concentration diagram of modified polyurethanes under molar ratio of [UPTI]: [Diamed-X] = 1:Y:

1. [CuCl₂]:[AGM]=1:4 (Y=0.9);
2. [CuCl₂]:[DEHA]:[AGM]=1:1,48:0,25 (Y = 0.9);
3. [CuCl₂]:[DEHA]:[AGM]=1:1,48:0,25 (Y = 0.7);
4. [CuCl₂]:[DEHA]:[AGM]=1:1,48:0,25 (Y = 0.5).

It was found that the use of metal complex systems based on CuCl₂, DEHA and AGM could significantly reduce the dosage of curing agent 4,4-methylene-bis-o-chloroaniline (Diamed-X) for urethane forming system based on UPTI.

7. The use of highly ordered coordination compounds of copper for receiving the rigid polyurethane foam

Metal complexes derived from CuCl₂, DEHA and AGM were also tested as modifiers of the polyol component used in the manufacturing of rigid polyurethane foam. It was found that metal complex system based on CuCl₂, DEHA and AGM had a significant impact on the rise and curing time of foam, reducing it (Fig. 12).

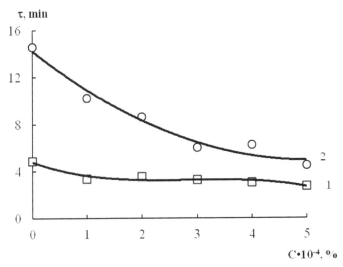

Figure 12. Rise (1) and curing (2) time of foam as a function of metal complex concentration based on [CuCl₂]:[DEHA]:[AGM]=1:1,48:0,25 in terms of CuCl₂ (%)

It was also established that the increasing of the molar ratio of the AGM in metal complex system led to even greater decrease in rise and curing time of foam (Fig. 13).

In order to establish the role of the AGM in the foaming process it was loaded alone in polyol component (Figure 14). It was found that the AGM also reduces the rise and curing time of the foam. However, these parameters were more than two times higher than the parameters that caused the addition of metal complex system.

We also used metal complex system based on CuCl₂ and DEHA as the control modifying system. In this case, in the wide range of concentrations of modifier the foam "collapsed". That is, the foam rose and the subsequently settled out. It should be also mentioned that the density of foam produced using metal complex system CuCl₂ - DEHA - AGM did not change in comparison with polyurethane foam obtained by the unmodified polyol component.

Thus, these studies show a significant catalytic effect of the metal complex modifier on foaming. In this connection it should be noted that the polyol component is a complicated balanced system that contains catalysts of amine nature and organotin compounds already. Our results suggest that the metal complex systems act as a cocatalyst.

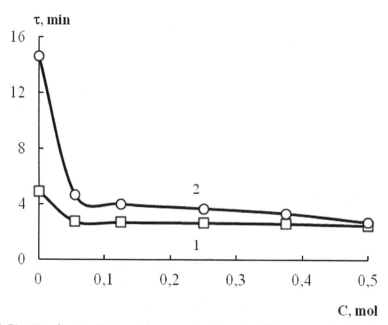

Figure 13. Rise (1) and curing (2) time of foam as a function of modifier concentration based on [CuCl₂]:[DEHA]:[AGM]=1:1,48:X, where X is a mole fraction of AGM in the metal complex overall concentration (mol)

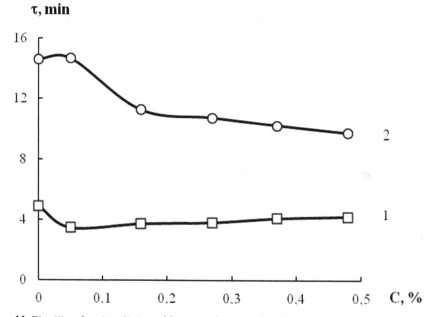

Figure 14. Rise (1) and curing (2) time of foam as a function of AGM concentration.

The next step was the research of such polyurethane foam key indicators as moisture absorption (Fig. 15) and water absorption (Fig. 16-17). It was found that the modified foam had enhanced characteristics as compared with unmodified polyurethane foam.

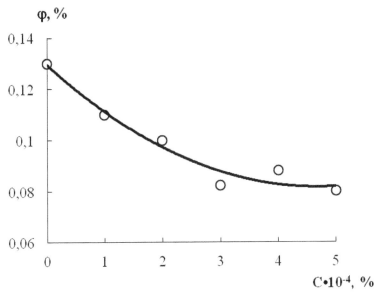

Figure 15. Moisture absorption of rigid foam as a function of metal complex concentration based on [CuCl₂]:[DEHA]:[AGM]=1:1,48:0,25 in terms of CuCl₂ (%)

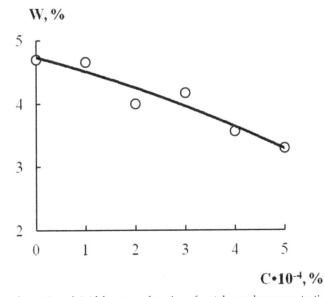

Figure 16. Water absorption of rigid foam as a function of metal complex concentration based on [CuCl₂]:[DEHA]:[AGM]=1:1,48:0,25, where in terms of CuCl₂ (%)

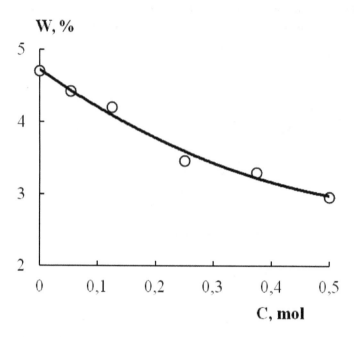

Figure 17. Water absorption of rigid foam as a function of modifier concentration based on [CuCl₂]:[DEHA]:[AGM]=1:1,48:X, where X is a mole fraction of AGM in the metal complex overall concentration (mol)

In conclusion, it should be noted that in order to achieve a positive result very small amounts of modifiers are required.

8. Conclusion

We considered the methods of obtaining transition metal coordination compounds that were active in reactions with isocyanate and urethane groups.

The feature of these metal complexes is that the metal ions are arranged in a chain of atoms linked together by chloride bridges. It is established that the chain of exchange-coupled transition metal ions remains in the polyurethanes structured by metal complex compounds. This circumstance is the cause of stepwise decrease in the specific volume resistivity of the modified polyurethanes.

It seems to be interesting for further research in this field to study the effect of metal complex binding on the physical and mechanical properties of polyurethanes. The most promising materials in terms of improving strength properties and heat resistance are thermoplastic urethanes.

Author details

Ruslan Davletbaev, Ilsiya Davletbaeva and Olesya Gumerova
Kazan National Research Technological University, Russia

9. References

Brostow W. (1990). Reaches of the liquid crystalline systems. *Polymer,* Vol.31, pp. (979-1023)

Carrher C.E. (1981). The structure of LC polymeric systems. *J. Chem. Ect.,* Vol.58, pp.(921-929)

Davletbaeva I.M., Kirpichnikov P.A. & Rakhmatullina A.P. (1996). Synthesis and investigation of liquid polyurethane metal complexes. *Macromolecular Symposia,* Vol.106, pp.(87-90)

Davletbaeva I.M, Shkodich V.F., Ismagilova A.I. & Parfenov V.V. (2001). Electro-physical properties of mesogenic metal-coordinated polyuretanes. *Russian polymer news,* Vol.6, No.4, pp.(36-38)

Dirk C.W., Bousseau M., Barret P.H., Moraes F., Wudl F. & Heeger A.J. (1986). Metal Poly(benzodithiolenes). *J. Macromolecules,* Vol.19, pp.(266-268)

Kingsborough R.P, Swager T.M. (1999). Polythiophene Hybrids of Transition-Metal Bis(salicylideninine)s: Correlation Between Structure and Electronic Properties. *J. Am. Chem. Soc.,* Vol.121, pp.(8825-8834)

Reynolds R., Karasz F.E., Lillya C.P. & Chien J.C.W. (1985). Electrically Conducting Transition Metal Complexes of Tetrathiooxalates. *J. Chem. Soc., Chem. Commun, pp.(268-269)*

Serrano J.-L., Oriol L. (1995). Metallomesogenic polymers. *J. Adv. Mater.,* Vol.7, No.4, pp.(348-369)

Shirai H., Vagi S., Suzuli A. (1977). Functional metal-porhyrazine derivatives and their polymers. 1. Synthesis of metal-phtalocyanine derivatives. *J. Macromol.Chem.,* Vol.178, No.7, pp.(1889-1895)

Shirai H., Kobayashi K. & Takemae V. (1979). Organometallic polymers. *J. Mac-romol.Chem.,* Vol.180, pp.(2073-2084)

Thuchide E.,Nishide H. (1977). Polymer-metal complexes and their catalytic activity. *Advances in Polymer Science,* Vol.24, pp.(2-87)

Wang F., Reynolds J.R. (1988). Soluble and electroactive nickel bis(dithiolene) complex polymers. *Macromolecules,* Vol.21, No.9, pp.(2887-2889)

Davletbaeva I.M., Ismagilova A.I, Tyut'ko K.A., Burmakina G.V. & Kuzaev A.I. (1998) Reactions of isocyanates with the system based on $CuCl_2$ - N,N'-Diethylhydroxylamine. *Russian Journal of General Chemistry,* Vol.68, No.6, pp.(1021-1027)

Davletbaeva I. M., Pyataev A. V., Kalachev K. E., Sadykov E. K. & Manapov R. A. (2006) Mössbauer study of structurally ordered iron coordination compounds and polyurethanes crosslinked by them. *Polymer Science,* Ser. A, 2006, Vol. 48, No. 6, pp.(612–617)

Chapter 3

New Liquid Crystalline Polyurethane Elastomers Containing Thiazolo [5,4d] Thiazole Moiety: Synthesis and Properties

Issam Ahmed Mohammed and Mohamed Rashidah Hamidi

Additional information is available at the end of the chapter

1. Introduction

Originally there are three states of matter; solid, liquid and gas. The emergence of an exotic and extraordinary form of matter, which is known as liquid crystal has been considered as one of the major breakthrough in polymer science. Liquid crystal can be defined as an intermediate of solid (crystal) and liquid (Knight & Vollrath, 2002) where the molecules have the capabilities to flow like a liquid (mobility) as well as possessing the common property associated to solid, which is the degree of order (Doldeny & Alder, 1998). In addition, liquid crystal materials are self assembling by nature and can offer a very elegant and effective way of controlling and tuning the physical properties that ultimately define the self-organizing and self assembly process (Zhang et.al., 2008). One of the exciting developments involving this unique material is the introduction of liquid crystalline behavior in polyurethane elastomers (PUE) where the first of this kind was synthesized by Iimura in 1981 (Lin et.al., 2001).

Polyurethane [PU] is one of the most versatile class of polymeric materials known today. Wide variety of structural changes can be produced with the different hydroxyl compounds and isocyanates leading to a wide spectrum of properties and applications (Yeganeh et.al., 2007). It contains a high concentration of polar groups, in particular the urethane group, resulting from isocyanate-hydroxyl reactions. The interactions between these polar entities are of great importance in determining the properties of PU of all types (Lee et.al., 1999) besides the composition and characteristic of the polyol, diisocyanates and the additives utilized during the synthesis (Pachecho et.al., 2009).

High toughness, excellent wear and tear properties and good oil resistance are among the advantages displayed by PUE (Wright & Cumming, 1969). Moreover, not only they have good mechanical and physical properties, PUE are also benefited with biocompatibility characteristics for biomedical applications (Barikani et.al., 2009). Despite all the great aforementioned properties, modifications and improvements are done to conventional PUE in order to meet the qualities in more advanced applications.

Diisocyanates, polyol and low molecular weight diamine or diol (chain extender) are the basic building blocks of conventional PUE (Yeganeh & Mehdizadeh, 2009). In order to synthesize liquid crystal polyurethane elastomers (LCPUE), the low molecular weight diamine or diol used in conventional PUE was substituted with the mesogenic unit. Incorporation of geometrically anisotropic moieties (mesogenic unit) within polymer architecture can drive the formation of liquid crystalline phase from strictly steric repulsion considerations (Abe & Ballauf, 1991; Rowan & Mather, 2007). Furthermore, the insertion of mesogenic unit in the backbone of PUE will impart unique physical properties to the polymer and also improve its mechanical, optical and electrical characteristics (Jia et.al., 1996).

Various mesomorphic behaviors are exhibited with different types of mesogenic units in preparation of LCPUE. In this research work, mesogens consumed were thiazolo [5.4d] thiazoles based and it is known as an important class of biycyclic aromatic molecule comprising two fused thiazole rings (Knighton et.al., 2010). Thiazolothiazole rigid fused ring can enhance the rigidity of the polymer and the conjugation (Osaka et.al, 2007) which makes it a best candidate to be part of the hard segment in the LCPUE network. The hard segments consisted of either 2,5-bis(4-hydroxyphenyl) thiazolo-[5,4d] thiazole or 2,5-bis(4-hydroxy-3-methoxyphenyl) thiazolo [5,4d] thiazole and 4,4'- methylene diphenyl diisocyanate (MDI). As for the soft segments, polyethylene glycol (PEG) 1000, 2000 and 3000 were involved.

The ultimate aim of this work is to synthesize new LCPUE with the presence of thiazolo-[5,4d] thiazole as a chain extender. Study and analysis were carried out to determine the effects and consequences of the introduction of thiazolo-[5,4d] thiazole moiety and the influence of various lengths of polyols on the properties of LCPUE.

2. Experimental

2.1. Materials

Vanillin and 4, 4'-methylene diphenyl diisocyanate (MDI) were purchased from Aldrich Co. (United States). Rubeanic Acid (dithiooxiamide) and 4-hydroxybenzaldehyde were obtained from MERCK Co. (Germany). Polyethylene glycol with molecular weight of 3000, 2000 and 1000 (PEG: Mn= 3000, 2000 and 1000) were purchased from Fluka Chemica (Switzerland). All the chemicals were utilized as received without any further purification. N,N-Dimethylformamide purchased from Aldrich (United States) was distilled over Calcium Hydride (CaH$_2$) through vacuum distillation before being used.

2.2. Synthesis of monomers and polymers

2.2.1. Synthesis of 2,5-bis(4-hydroxyphenyl)thiazolo-[5,4d] thiazole (I)

Briefly, 3 g (25 mmol) of dithiooxamide (Rubeanic acid) and 15 g (123 mmol) of 4-hydroxybenzaldehyde with the presence of 9 g (97 mmol) of phenol were charged all at once in a 500 ml round bottom flask fitted with condenser and left to be refluxed for 2h. Precipitates were obtained by pouring the hot mixtures to the cold water. Subsequently, the yield was filtered off and washed with ethanol followed by ether. The product obtained was dried at 70°C in a vacuum oven for 24 hours. Recrystallization from cyclohexanone was performed giving an orange-yellowish powder. Yield: 35% with melting point 364°C. Fourier transform infrared (FTIR; KBr, cm^{-1}): 3492 (-OH), 1606 (C=N), 1596 (C=C), 855 (p-substituted benzene). ^{1}H-NMR (400 MHz, DMSO-d$_6$ ppm): δ_H 7.12 (m, aromatic protons), 9.8 (s, -OH). Elemental analysis: Found: C, 59.16; H, 3.28; N, 8.84, C$_{16}$H$_{10}$N$_2$O$_2$S$_2$ Calc.: C, 58.89; H, 3.09, N, 8.59.

2.2.2. Synthesis of 2,5-bis(4-hydroxy-3-methoxyphenyl) thiazolo [5,4d] thiazole (II)

The same procedure was applied to the synthesis of 2,5-bis(4-hydroxy-3-methoxyphenyl) thiazolo [5,4d] thiazole except that 4-hydroxybenzaldehyde was substituted with vanillin. Orange-yellowish powder was obtained as the end product. Yield: 26% with the melting point of 259°C. Fourier transform infrared (FT-IR; KBr disc): 3534 cm^{-1} (OH), 1608 cm^{-1}(C=N), 1510 cm^{-1} (C=C), 842 cm^{-1} (-CH out of plane). ^{1}H-NMR (400 MHz, DMSO-d$_6$ ppm): δ_H 7.09 (m, aromatic protons), 9.5 (s, -OH) and 3.87 (s, OCH$_3$). Elemental analysis: Found: C, 55.60; H, 4.03; N, 6.89, C$_{18}$H$_{14}$N$_2$O$_4$S$_2$ Calc.: C, 55.95; H, 3.62, N, 7.25.

2.2.3. Synthesis of liquid crystalline polyurethane elastomers (LCPUE)

Preparation of LCPUE was achieved by two steps solution polymerization reaction, where isocyanate terminated pre-polymer was synthesized initially in the first stage. To produce pre-polymer, 0.01 mol of PEG (Mw = 1000, 2000, and 3000) and 0.02 mol of MDI were mixed in 500ml of reactor flask equipped with condenser, thermometer, nitrogen inlet and mechanical stirrer. The mixture was allowed to be stirred and heated for 4h at 70°C in the presence of 15 ml of DMF as solvent and nitrogen gas was kept flowing to provide inert atmosphere. The reaction was followed by chain extension process, using either compound (I) or (II), where the chain extender was added dropwise within 1h to complete the formation of LCPUE. Subsequently, the temperature was increased to 100°C and the reaction continued for another 9 hr. The hot viscous solution was then poured into 200ml of cold water for precipitation, before subjected to filtration. Later, the filtered product was washed with ethanol several times and finally with ether, before being dried overnight in a vacuum oven at 60° C.

2.4. Measurements

100mg mixture of samples and KBr (grounded) were pressed into translucent disc before being subjected to Nicolar Avatar Model 360 Fourier Transform infrared spectrometer devices to obtain FT-IR spectra. Data was collected in the range of 4000-400cm^{-1}. ^{1}H-NMR and ^{13}C-NMR spectra were obtained using Bruker 400 MHz NMR spectrometer consuming DMSO-d$_6$ as solvent and TMS as internal standard. Thermal stability of LCPUE was determined by thermogravimetric analyzer (Perkin Elmer Pyris series 6) under nitrogen purge and with 10°C/min of heating rate and the heating was done up to 800°C. Liquid crystalline behavior was verified by means of differential scanning calorimetry (DSC) to observe the behavior of polymers such as glass transition point (T_g), melting point (T_m) and isotropic temperature (T_i). It was conducted utilizing Perkin Elmer Pyris Series 7 thermal analyzer under Nitrogen flux at 10^0C/min rate of heating. Textures of mesomorphic phases were displayed by Nikon Eclipse E600 polarized microscope equipped with MS600 Linkam Hot stage and SONY CCD-IRIS Color Video Camera. The heating rate was 5°C/min and 10°C/min for the cooling rate. Sample was placed between two thin round glasses and it was then transferred onto microscope fitted with the hot stage to be analyzed. Siemens X-ray Diffractometer model D5000 equipped with a CuKα target at 40KV and 40mA was used in obtaining X-ray scattering curve. Tensile strain properties of LCPUE films were measured by Instron Testing instrument at a constant speed of 500mm/min (speed) where the measurements were performed at room temperature. Brookfield viscometer was used to measure the fluid viscosity where suitable spindle and speed were chosen and it was also performed at room temperature.

3. Results and discussion

3.1. Preparation of chain extender

The preparation of 2,5-bis(4-hydroxyphenyl)thiazolo-[5,4d] thiazole and 2,5-bis(4-hydroxy-3-methoxyphenyl) thiazolo [5,4d] thiazole were conducted according to the reaction shown in Scheme 1. The starting reagent involved for the synthesis of both the compounds were rubeanic acid and either 4-hydroxybenzaldehyde or vanillin with the presence of phenol. Subsequently, both chain extenders prepared were being used in the preparation of LCPUE. Identification of the chemical structures of the aforementioned products was monitored primarily with FT-IR spectroscopy and further confirmation was carried out by ^{1}H-NMR spectrophotometer.

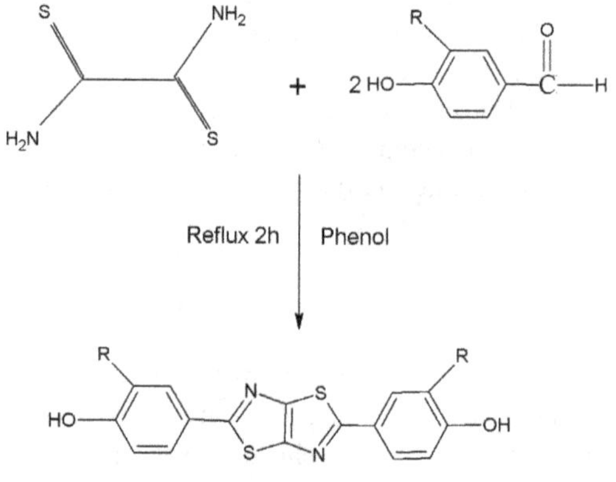

(I or II)

where R = -H (I) or -OCH$_3$ (II)

Scheme 1. Preparation of Compound I and II

3.2. Polymer synthesis

LCPUE based on thiazolo [5,4d] thiazoles moiety were synthesized from long chain of diol (PEG 3000, 2000 and 1000) with an excess of diisocyanate (MDI) via addition reaction to give the terminal reactive group which results in the formation of 'extended diisocyanate' or isocyanate pre-polymer. Then, 2,5-bis(4-hydroxyphenyl) thiazolo-[5,4d] thiazole [I] and 2,5-bis(4-hydroxy-3- methoxyphenyl) thiazolo [5,4d] thiazole [II] were added acting as a chain extender in order to convert the pre-polymer into long chain LCPUE. The general route for the preparation of LCPUE was outlined in Scheme 2, yield and viscosity of LCPUE were listed in Table 1 and the data showed that the range of the viscosities and yields obtained were 10,744 to 40 692 cP and 76-87 %, respectively. Range of the viscosities obtained also provides the information of the molecular weight of each polymer synthesized where high value of viscosity indicates high molecular weight of the polymer produced and vice versa (Bagheri & Pourmoazzen, 2008). In this case, all LCPUE samples displayed fairly high molecular weight in accordance with the results demonstrated.

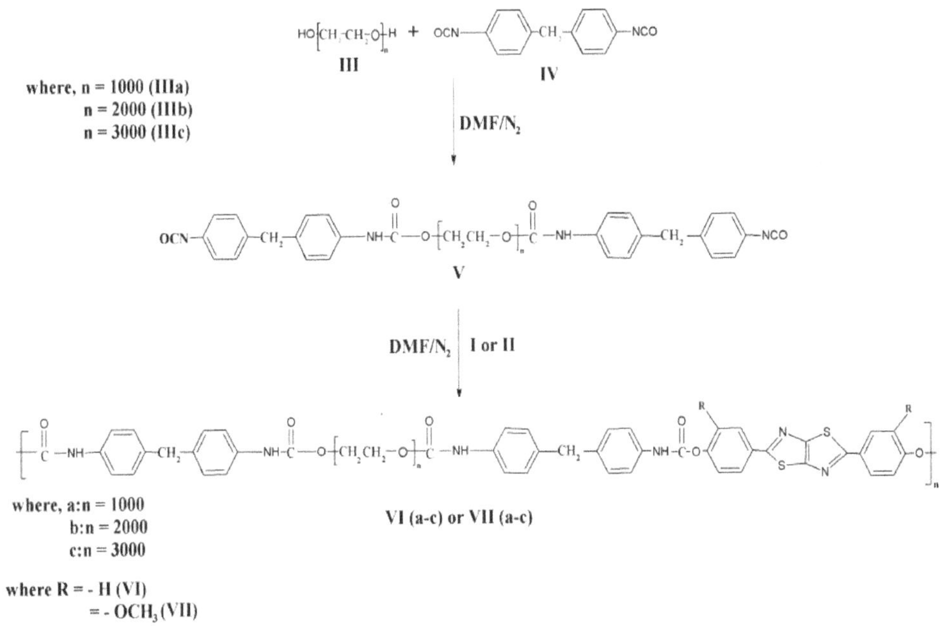

Scheme 2. General route for the preparation of LCPUE VI (a-c) and VII (a-c)

3.3. Structural elucidation

FT-IR was employed to verify functional groups of the pre-polymer, compound I and II, and LCPUE. Prior to the formation of LCPUE which is referring to the pre-polymer state, in the region of 2270 cm^{-1} a peak was observed which was assigned to –N=C=O-

(diisocyanate) whereas according to the IR spectra of compound I and II, a peak was found at 3492 cm⁻¹ and 3334 cm⁻¹ which corresponds to –OH functional group in the chemical structure. The disappearance of both the bands of -N=C=O- in pre-polymer and – OH of compound I and II, indicates the completion of the reaction of preparation of LCPUE and this fact was also supported with the appearance of new absorption bands at 3356.84cm⁻¹ (N-H- stretching) and 1782.5cm⁻¹ (carbonyl group) which were attributed to the urethane linkage, –NHCOO- (Zhang et al., 2008; Issam, 2007). Furthermore, the peak at 2884.89 cm⁻¹ was ascribed to –CH stretching, whereas the band representing C=C aromatic can be found at 1598.59 cm⁻¹. Figure 1 displayed the FTI-R spectrum of LCPUE VIIa and based on the results obtained, the characteristic absorption bands of FT-IR spectra for the other LCPUE were almost identical to one another. The fact that differentiates LCPUE VI and LCPUE VII was the presence of the methoxy group and it was proven in the FTIR spectrum of LCPUE VIIa, where a peak displayed at the region of 1024.27 cm⁻¹ corresponded to the methoxy group.

Further confirmation of chemical composition of LCPUE produced was carried out by means of Nuclear Magnetic Resonance spectroscopy (NMR). ¹H-NMR spectrum of LCPUE VIIa was illustrated in Figure.2. A singlet peak centered at 8.76 ppm was assigned to – NHCOO- and this proved the formation of urethane linkage. The appearance of multiplet peaks at 7.53-6.99 ppm and singlet peak at 3.87 ppm was attributed to the aromatic protons and the protons in methoxy group, respectively. Aliphatic chain of polyol (PEG 1000) was detected in the region of 1.23-1.64 ppm.

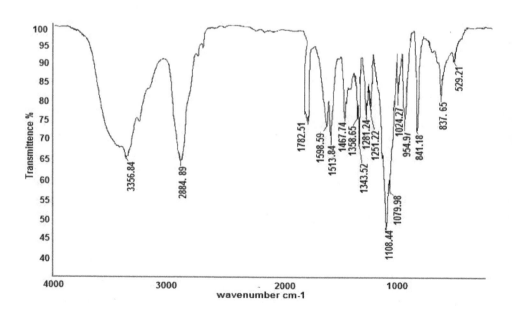

Figure 1. FTIR spectrum of LCPUE VIIa

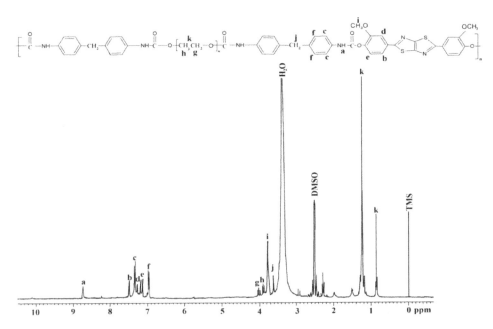

Figure 2. ¹H-NMR spectrum of LCPUE VIIa

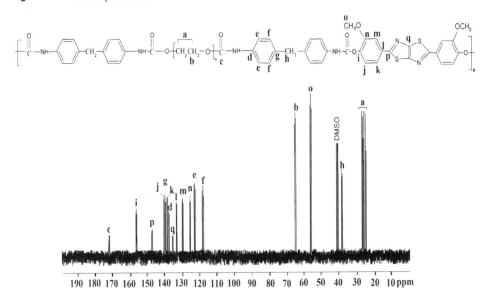

Figure 3. ¹³ C-NMR spectra of LCPUE VIIa

Other than FT-IR and ¹H-NMR analysis, ¹³C-NMR was performed in order to clarify the structure of LCPUE prepared. ¹³C-NMR spectra portrayed in Figure.3 which represents LCPUE VIIa shows that the formation of urethane linkage (NHCOO) was determined by the appearance of the peak at 173.4 ppm. The methylene group presence in the soft segment of PEG can be seen as a sharp and intense peak at 25-29 ppm. More peaks can be observed at 117.8 to 158.7 ppm and 56.2 ppm where they were assigned to the aromatic carbons and the carbon in methoxy group respectively. Significant peaks in all characterization analysis (FT-IR, ^{1}H-NMR and ^{13}C-NMR) were consistent and adequately provide the evidences to support the fact that the reaction of all materials took place and LCPUE was successfully prepared.

3.4. Thermal and liquid crystalline behavior of polymers

The DSC analysis was conducted at a heating rate of 10°C to understand phase separation behavior of all synthesized LPCUE where the transition occurs, observed under polarizing optical microscope (POM) equipped with heating stage and the results obtained from both measurements were listed in Table 1. Based on the DSC thermograms, upon heating, one step transition and two endothermic peaks were detected where each of them indicates glass transition (T_g), melting endotherm, (T_m) and isotropic endotherm (T_i) respectively, which is also the evidence of the existence of mesophase. LCPUE derived from 2,5-bis(4-hydroxy-3-methoxyphenyl) thiazolo [5,4d] thiazole have transition temperatures lower than those derived from 2,5-bis(4-hydroxyphenyl)thiazolo-[5,4d]thiazole. Methoxy group, which acts as a substituent attached to the phenyl ring has the capability to lower the melting and isotropization temperature and caused thermal suppression of the molecule to occur (Al-Dujaili et.al., 2001). The fact was supported by the results illustrated in Fig.4 where it depicts the DSC thermograms of LCPUE. LCPUE VIIa displayed melting point (T_m) at 164°C and isotropization temperature (T_i) at 187°C whereas for LCPUE VIa, T_m was detected at 176°C and T_i at 205°C. The substituent could also act to reduce the coplanarity of adjacent mesogenic groups and increase the diameter or decrease the axial ratio of the mesogens [Li and Chang, 1991]. Due to the higher range between T_m and T_i of LCPUE VIa, the thermal properties of this polymer are higher and more stable compared to LCPUE VIIa. The types of diisocyanates also contribute to the thermal behavior of LCPUE, where MDI based PU was known for having better order of the rigid chain that approaches the decomposition temperature, giving high melting point to the polymer produced (Jieh & Chou, 1996). As for glass transition, it involves mobility of the chain segments and the Tg will be affected by the mobility restriction on the chain segments, (Suresh et.al., 2008) it therefore explains the varying pattern of the T_g values displayed in Table 1. The decreasing values of T_g can be observed as the length of soft segments increases, indicating that the long chain of polyol gave great flexibility characteristics towards the polymer chains where less mobility restrictions occurred and hence resulting in the lower T_g values.

POM was utilized to investigate the type of mesophase by displaying the phase transition that occurred, subsequently providing the polarizing optical microphotographs of the target compounds. The morphology observed on heating and transition temperatures obtained were given in Figures 5 and 6 and the results were summarized in Table 1. It was revealed that all LCPUE showed mesophases upon melting temperature where the thread texture of the nematic phases can be seen. From the photographs taken by POM, the crystal to mesophase transition occurred at temperature ranging from 129 to 181°C. The samples were further heated after the crystal-nematic transition temperature, and resulted in the disappearing of the texture when reaching the isotropization stage. There were no traces of mesophase transition during the cooling process from POM indicating all samples possessed thermotropic type of liquid crystal. Phase transition temperatures observed through POM were found to be consistent with the corresponding DSC thermograms.

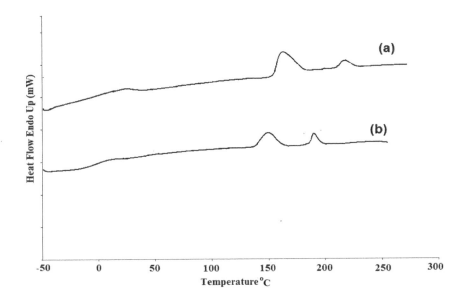

Figure 4. DSC traces of (a) LCPUE VIa (b) LCPUE VIIa

SAMPLE	PEG MOLECULAR WEIGHT	Yield (%)	Viscosity cP	DSC			POM	
				T_g	T_m	T_i	T_m	T_i
				(°C)	(°C)	(°C)	(°C)	(°C)
LCPUE VIa	1000	85	11 108	25.1	176	205	181	200
LCPUE VIb	2000	83	26 456	22.5	153	174	162	180
LCPUE VIc	3000	77	40 692	19.1	139	156	133	161
LCPUE VIIa	1000	76	10 744	15.2	164	187	170	193
LCPUE VIIb	2000	80	22 453	11.8	143	163	148	170
LCPUE VIIc	3000	87	39 981	10.4	125	142	129	149

Table 1. Thermal properties of LCPUE VI (a-c) and LCPUE VII (a-c) by DSC and POM

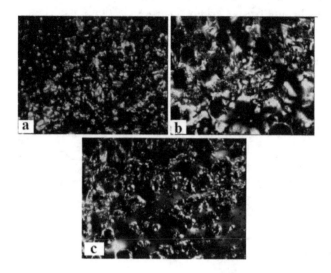

Figure 5. Polarized optical images of (a) LCPUE VIa (181°C), (b) LCPUE VIb (162 °C) and (c) LCPUE VIc (133 °C)

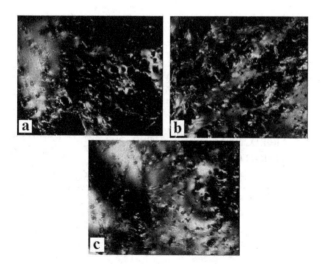

Figure 6. Polarized optical images of (a) LCPUE VIIa (170 °C), (b) LCPUE VIIb (148 °C) and (c) LCPUE VIIc (129 °C)

X-ray diffraction analysis of LCPUE was conducted at room temperature to obtain information on both the mesophase structure and crystallinity of LCPUE. The measurements exhibited several peaks in the range of $2\theta= 15 - 25°$ as observed in Figure 7 and this indicated semi crystalline character possessed by LCPUE. The results obtained in above range also provide details related to the d-spacing of 3.56 and 4.92 Å, thus supporting the characteristic of nematic liquid crystalline phase (Jeh & The, 1994) as displayed through POM.

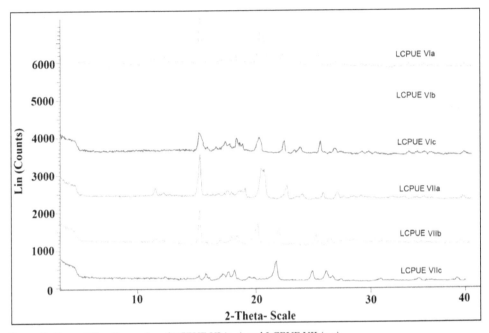

Figure 7. X-ray diffraction scales of LCPUE VI (a-c) and LCPUE VII (a-c)

Thermal stability of prepared LCPUE was investigated by thermogravimetric analysis (TGA). Incorporation of liquid crystalline properties into the polymer structure would enhance the thermal properties (Jahromi et.al., 1994) and the theory has proved to be applicable from the results obtained. This may be partly due to favorable interactions between hard domain interface and the liquid crystalline phase. All synthesized LCPUE possessed good thermal stabilities, however, PU elastomers eventually undergo thermal degradation when exposed to high temperatures. Degradation process occurred in two step pattern where the initial degradation occurs in the hard segment involving the urethane linkages, while the second stage indicated the degradation of soft segments. TGA curves in Figure 8 demonstrated the thermal degradation of all LCPUE prepared where 10% weight loss of LCPUE occurred at about 315-341°C and the maximum degradation temperature was in the range of 430-470°C, signifying a high thermal stability property. Furthermore, it can be observed that LCPUE VIIc demonstrated the lowest degradation temperature among the others and this proved that the length of polyethylene glycol (soft segment) influenced the thermal stability of LCPUE where the order of LCPUE due to their thermal stability can be arranged as LCPUE VIa>VIIa>VIb>VIIb>VIc>VIIc.

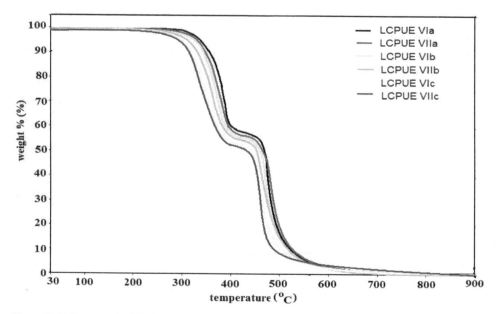

Figure 8. TGA curve of LCPUE VI (a-c) and LCPUE VII (a-c)

3.5. Tensile properties

Table 2 demonstrates tensile properties of the synthesized LCPUE. As seen, all of the polymers possessed good elastic properties with high elongation at break. Due to the data listed, the higher the molecular weight of the soft segments, the greater the elongation at break, but decrease of tensile strength and tensile modulus can be observed. When the molecular weight of polyol increased, the number of urethane groups in the polyol chain was reduced at the same time, and hence the number of rigid segments is lower, consequently, the possible number of intermolecular hydrogen bonds goes down in which –NH and C=O groups are active (Kro & Pitera, 2008). However, the presence of enhanced rigid and high aspect ratio mesogenic unit as part of hard segment in the synthesized LCPUE, is able to give both high strength and good elastic properties to LCPUE even with long soft segments, which

Sample	Tensile modulus (Mpa)	Tensile strength (Mpa)	Elongation at break (%)
LCPUE VIa	17.1	28.2	290
LCPUE VIb	13.4	24.1	450
LCPUE VIc	11.5	19.9	570
LCPUE VIIa	17.5	28.3	330
LCPUE VIIb	13.7	24.3	460
LCPUE VIIc	11.2	19.8	560

Table 2. Mechanical properties of LCPUE VI (a-c) and LCPUE VII (a-c)

is unusual in conventional PUE (Jeong et.al., 2000). Better phase separation will lead to good mechanical properties; hence the introduction of the mesogens unit as chain extender into LCPUE can be said to easily induce the matter (phase separation) to occur.

Author details

Mohammed Ahmed Issam and Hamidi Mohamed Rashidah
University Sains Malaysia, Malaysia

Acknowledgement

The author would like to thank University Sains Malaysia for short term grant no.304.PTEKIND.6311031 and the fellowship scheme for funding the research.

4. References

Abe, A. & Ballauf, M. (1991). Liquid crystallinity in Polymers. John Wiley & Sons Inc, New York, USA

Al-Dujaili, A.H.; Atto, A.T. & Al-Kurde, A.M. (2001). Synthesis and Liquid Crystalline Properties of Models and Polymers containing Thiazolo[5,4-d]thiazole and Siloxane Flexible Spacers. European Polymer Journal Vol.37, pp. 927-932

Bagheri, M. & Pourmoazzen, Z. (2008). Synthesis and Properties of New Liquid Crystalline Polyurethanes containing Mesogenic Side Chain Reactive & functional. Polymers, Vol.68, pp. 507–518

Barikani, M.; Honarkar, H. & Barikani, M. (2009). Synthesis and Characterization of Polyurethane Elastomers based on Chitosan and Poly(e-caprolactone). Journal of Applied Polymer Science, Vol.112, pp. 3157–3165

Doldeny, J.D. & Alder, P.T. (1998). The Mesogenic Index: An Empirical Method for Predicting Polymeric Liquid Crystallinity. High Performance Polymers, Vol.10, pp. 249–272

Issam, A.M. (2007). Synthesis of Novel Y-Type Polyurethane containing Azomethine Moiety, as Non-linear Optical Chromophore and Their Properties. European Polymer Journal, Vol.43, pp. 214-219.

Jahromi, S.; Lub, J. & Mol, G.N. (1994). Synthesis and Photoinitiated Polymerization of Liquid Crystalline Diepoxides. Polymer, Vol. 35, No.3, pp. 622-629

Jeh, C.T. & Teh, C.C. (1994). Study on Thermotropic Liquid Crystalline Polymers -I. Synthesis and Properties of Poly(azomethine-urethane)s. European Polymer Journal, Vol.30, pp. 1059-1064

Jeong, H.M.; Kim, B.K. & Choi, Y.J. (2000). Synthesis and Properties of Thermotropic Liquid Crystalline Polyurethane Elastomers. Polymer,Vol.41, pp. 1849-185

Jia, X.; He, X.D. & Yu, X.H. (1996). Synthesis and Properties of Main-Chain liquid Crystalline Polyurethane Elastomers with Azoxybenzene. Journal of Applied Polymer Science, Vol.62, pp. 465-47

Jieh, S.S. & Chou, C.T. (1996). Studies on Thermotropic Liquid Crystalline Polyurethanes.III.Synthesis and properties of polyurethane elastomers by using various mesogenic units as chain extender. Journal of polymer science part A: Polymer chemistry, Vol.34, pp. 771-779

Knight, D.P. & Vollrath, F. (2002). Biological Liquid Crystal elastomers. Philosophical Transactional Royal Society London. B, Vol.357, pp. 155–163

Knighton, R.C.; Hallett, A.J.; Kariuki, B.M. & Pope, S.J.A. (2010). A One-step Synthesis towards New Ligands based on Aryl-functionalized Thiazolo[5,4-d]thiazole Chromophores. Tetrahedron Letters, Vol.51, pp. 5419–5422

Kro, P. & Pitera, B.P. (2008). Mechanical Properties of Crosslinked Polyurethane Elastomers Based on Well-Defined Prepolymers. Journal of Applied Polymer Science, Vol.107, No.3, pp. 1439–1448

Lee, D.J.; You, S.H. & Kim, H.D. (1999). Synthesis and properties of thermotropic liquid crystalline polyurethane elastomers (II): Effect of Structure of Chain Extender Containing Imide Unit. Korea Polymer Journal, Vol.7, No.6, pp. 356-363

Lin, C.K.; Kuo, J.F. & Chen, C.Y. (2001). Synthesis and Properties of Novel Polyurethanes containing the Mesogenic Moiety of a-Methylstilbene Derivatives. European Polymer Journal, Vol.37, pp. 303-313

Li, C.H.; & Chang, T.C. (1991). Thermotropic Liquid Crystalline Polymer:III: Synthesis and Properties of Poly(amide-azomethine-ester). Journal of Polymer Science Part A: Polymer Chemistry, Vol.29(3), pp. 361-367

Osaka, I.; Sauvé, G.; Zhang, R.; Kowalewski, T. & McCullough, R.D. (2007). Novel Thiophene-Thiazolothiazole Copolymers for Organic Field-Effect Transistors. Advance Material, Vol.19, pp. 4160–4165

Pacheco, M.F.M.; Bianchi, O.; Fiorio, R.; Zattera, A.J.; Giovanel, M.Z.M. & Crespo, J.S. (2009). Thermal, Chemical, and Morphological Characterization of Microcellular Polyurethane Elastomers. Journal of Elastomers and Plastics, Vol.41, pp. 323

Rowan, S.J. & Mather, P.T. (2008). Supramolecular Interactions in the Formation of Thermotropic Liquid Crystalline Polymers. Structure and bonding, Vol.128, pp. 119-149

Suresh, K.I.; Tamboli, J.R.; Rao, B.S.; Verma, S. & Unnikrishnan G. (2008). Effect of Core Group Substituents on the Monomer Mesophase, Photocuring, and Film Viscoelastic Properties of Mesogenic Diacrylates. Polymers for Advanced Technologies, Vol.19, pp.1323-1333

Wright, P. & Cumming, A.P.C. (1969). Solid Polyurethane elastomers. Mclaren and Sons, London

Yeganeh, H. & Mehdizadeh, M.R. (2004). Synthesis and Properties of Isocyanate Curable Millable Polyurethane Elastomers Based on Castor Oil as a Renewable Resource Polyol. European Polymer Journal, Vol.40, pp. 1233–1238

Yeganeh, H.; Talemi, P.H. & Jamshidi, S. (2007). Novel Method for Preparation of Polyurethane Elastomers with Improved Thermal Stability and Electrical Insulating Properties. Journal of Applied Polymer Science, Vol.103, pp. 1776–1785

Zhang, C.; He, Z .; Wang, J.; Wang, Y. & Ye, S. (2008). Study of Mesogenic Properties and Molecular Conformation from a Heterogeneous Tetramer with a Triphenylene Centre Core and Three Cyanobiphenyl Tails. Journal of Molecular Liquids, Vol. 138, pp. 93–99

Zhang, H.; Chen, Y.; Zhang, Y.; Sun, X.; Ye, H. & Li, W. (2008). Synthesis and Characterization of Polyurethane Elastomers. Journal of Elastomers and Plastics, Vol.40, No.2, pp. 161-177

Bottom-Up Nanostructured Segmented Polyurethanes with Immobilized *in situ* Transition and Rare-Earth Metal Chelate Compounds – Polymer Topology – Structure and Properties Relationship

Nataly Kozak and Eugenia Lobko

Additional information is available at the end of the chapter

1. Introduction

The formation of the polyurethanes (PU) with immobilized *in situ* co-ordinating metal compounds allows obtain structurally homogeneous systems with uniform dispersed nanosize metal containing sites. Aggregation of these metal chelate compounds is prevented due to complexing with polar groups of the polymer matrix.

At the same time due to complex formation between the metal compound and polymer functional groups, the structuring of the forming matrix occurs on a nanoscale level. As a result, in the presence of small amounts of metal chelate compound (0,5-5%wt) both change of the polyurethane structure and properties can be observed. To understand the nature of above phenomena the influence of the weak interactions «macromolecule - metal» were analyzed on the metal-containing PUs structure, molecular dynamics and properties.

The present study investigates the formation of nanostructured linear and cross-linked polyurethanes (LPUs and CPUs, respectively) with immobilized *in situ* mono- and poly-heteronuclear chelate compounds of rare-earth and transition metals. Influence of PU topology on self organization processes in polymer matrix and its properties is also subject of analysis.

1.1. Materials, methods and instrumentations

Polypropylene glycol (PPG, MW 1000) was dried under vacuum at 120 ºC for 2 h. Tolylene diisocyanate (mixture 80/20 of 2,4- and 2,6- isomers) (TDI) was distilled under vacuum.

Diethylene glycol (DEG) was distilled under vacuum at 105 °C. Trimethylol propane (98%) (TMP) was dried under vacuum at 40-45 °C for 2-4 h. Dichloromethane (CH_2Cl_2), 1,4-dioxane and N,N'- dimethylformamide (DMF) were distilled at 40 °C, 101 °C, 153 °C, respectively. The following chelate compounds of transition and rare-earth metals as PU modifier were used:

		(R=-CH₃) Cu(acac)₂ - Copper(2+) acetylacetonate Ni(acac)₂ – Nickel(2+) acetylacetonate (R = -OC₂H₅) Cu(eacac)₂ - Copper (2+) ethyl acetoacetate (R = --CF₃). Cu(tfacac)₂ - Copper (2+)trifluoro acetylacetonate
		(R₁=R₂=-CH₃) Co(acac)₃ – Cobalt (3+) acetylacetonate Cr(acac)₃ – Chromium (3+) acetylacetonate Gd(acac)₃ – Gadolinium (3+) acetylacetonate Nd(acac)₃ – Neodymium (3+) acetylacetonate Er(acac)₃ – Erbium (3+) acetylacetonate (R₁ =-C(CH₃)₃ , R₂ =-(CF₂)₂-CF₃) Eu(fod)₃ – Europium (3+) tris(6,6,7,7,8,8,8-heptafluoro-2,2-dimethyl-3,5-octanedione) (R₁ =- thiophene , R₂ =- CF₃) Eu(TTA)₃ – Europium (3+) thenoyltrifluoroacetonate
		(R₁ =- thiophene , R₂ =- CF₃ ; L₄= phen) Eu(TTA)₃ phen – Europium (3+) tris(thenoyltrifluoroacetonate) phenantroline (R₁ =- thiophene , R₂ =- CF₃ ; L₄= triphenylphosphine oxide) Eu(TTA)₃ TPPO –Europium (3+) tris (thenoyltrifluoroacetonate) (triphenylphosphine oxide)
$\left[Met_k^1 Met_m^2 Met_n^3 R_p^1 R_q^2 R_r^3 \right] \cdot Sol_t$ $p,q,k,m = 1,2,3,4;$ $n,r,t = 0,1;$ where Me₂Ea = deprotonated residue of dimethyl aminoethanol Dea = doubly deprotonated residue of diethanolamine		k=2, m=1, n=0, p=3, q=3, r=0, t=1, R¹=NCS, R²=Me₂Ea, Sol= CH₃CN [Cu₂Zn(NCS)₃(Me₂Ea)₃].CH₃CN
		k=2, m=3,n=0, p=6, q=4, r=0, t=2, R¹=Br, R²=Me₂Ea, Sol= dmso [Cd₂Cu₃Br₆(Me₂Ea)₄(dmso)₂]
		k=1, m=2, n=2, p=3, q=4, r=4, t=0, R¹= H₂Dea, R²= NCS, R³= Dea [Ni(H²Dea)²][CoCu(Dea)²(H²Dea)(NCS)]²(NCS)²

Table 1. The PU modified chelate compounds of transition and rare-earth metals.

In metal chelate compounds used as PU modifier metal ions are already surrounded with organic ligands. This facilitates solvation of modifier in polymer. The listed above transition and rare-earth metal chelate compounds are commercial products (Aldrich). The heteroligand rare-earth metal compounds were synthesized by Professor Svetlana B. Meshkova's group (A. V. Bogatsky Physic-Chemical Institute of National Academy of Sciences of Ukraine, Odessa). Polyheteronuclear metal complexes of Cu (2+), Cd (2+), Zn (2+), Ni (2+) and Co (3+), described in (Skopenko et al., 1997; Vinogradova et al., 2002), were provided by Prof. V. Kokozay's goup (Kiev Taras Shevchenko University). Polyheteronuclear metal chelate compounds can realize unexpected coordination states of transition metal ions. That, in turn, can give new properties to a polymer formed in their presence.

PUs were synthesized in two stages according to standard procedure described in detail elsewhere (Saunders&Frish,1968; Wirspza,1993) using PPG-1000 and TDI based prepolymer. DEG was used as chain extender to obtain LPU (Scheme 2). TMP was used as cross-linking agent to obtain CPU (Scheme 3). Metal chelate compounds were added into reaction mixture as solution in CH_2Cl_2, 1,4-dioxane or DMF to obtain the metal containing PUs with homogeneous distribution of modifier (from 0,5 to 5 %wt.) in polymer matrix. High ability of metal chelate compound to complex formation leads to enrichment of PU matrix with heteroligand macro complexes of 3d- and 4f-metal with prevalence of outer-sphere coordination of macro chains. Such macro complexes act like coordination linkages between polymer chains and form "coordination nodes" in PU (Scheme 1).

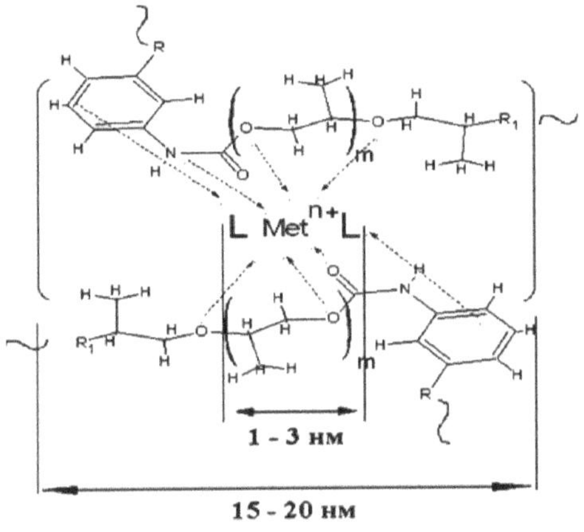

Scheme 1. The coordination junction of PUs networks.

Thus, in the LPU (Scheme 2) in the presence of chelate metal compounds the "coordination nodes" can form.

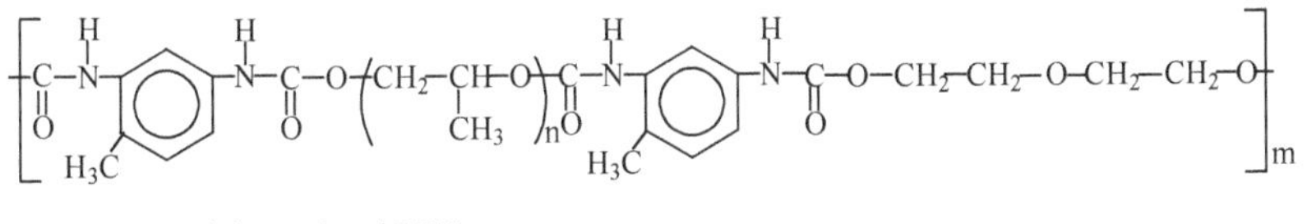

Scheme 2. The general formula of LPU.

In the metal containing CPU both the chemical linkages (Scheme 3) and the "coordination nodes" can form (Scheme 1).

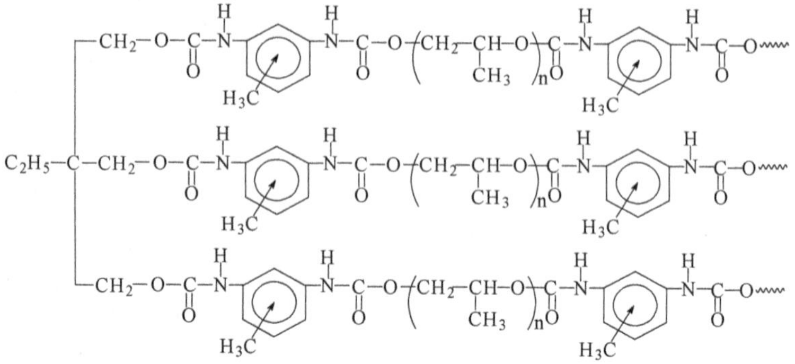

Scheme 3. The fragment of PU network with cross-linkage.

Wide-angle X-ray scattering (WAXS) profiles of studied samples were recorded on a Dron-4-07 diffractometer with Ni-filtered Cu-K_α radiation and Debay-Sherer optical schema. Distance between PU atomic layers (d) was estimated using the Bragg equation:

$$\lambda = 2d\sin\theta \qquad (1)$$

where λ – the X-ray wave length (λ = 0,154 nm); θ - the diffraction maximum angular position, degrees.

Small-angle X-ray scattering (SAXS) profiles were recorded using KPM-1 X-ray camera (Kratky et al., 1966). The Schmidt's method (Schmidt & Hight, 1960) was used to smooth out the SAXS-profiles to point collimation. X-ray measurements are carried out using monochromatic Ni-filter of Cu-K_α radiation at temperature 22±2 °C. The Bragg's period of uniform electronic density scattering elements was estimated through the equation:

$$D = 2\pi / q \qquad (2)$$

The X-band EPR-spectra were recorded at temperature 20°C using radio spectrometer PE-1306 equipped with frequency meter ChZ-54. The magnetic field was calibrated using 2, 2-diphenil-1-pycrilhydrazyl (DPPH) (g=2,0036) and ions of Mn(2+) in MgO matrix (g=2,0015).

Stable nitroxide radical 2,2,6,6-tetramethylpiperidinyl-1-oxy (TEMPO) was used as paramagnetic spin probe (SP). Nitroxide SP was introduced into PU films via diffusion of its saturated vapor at 30°C for 2 hours with subsequent keeping at 20°C for 24 hours.

Correlation time (τ) of SP rotational diffusion in the range of its fast motion ($10^{-11} < \tau < 10^{-9}$s) was calculated according (Vasserman & Kovarskii, 1986) as follows:

$$\tau = 6,65\Delta H_{(+1)}\left(\sqrt{(I_{+1}/I_{-1})}-1\right)\times 10^{-10}c, \qquad (3)$$

where $\Delta H_{(+1)}$ – is width of the low-field- component of TEMPO EPR-spectrum, I_{+1} and I_{-1} - are intensities of low-field and high-field components of the spectrum, respectively.

The differential scanning calorimetry in temperature interval from 223 to 750 K was performed using Perkin Elmer DSC 2 instrument with the IFA GmbH's software. The heating rate was 0,05-2 grad/min.

Micro images in light transmission were obtained using an optical microscope XY-B2 (NS Instr. Co.) equipped with digital video ocular ICM 532 and AMCAM/VIDCAP (Microsoft) image processing system.

The surface tension of PUs (γ_{sg}) was determined according to Elton's equation (Tavana et al., 2004) using measurement of contact wetting angle with ethyleneglycol (EG) as wetting liquid at 20°C:

$$\gamma_{sg} = 0,5\gamma_{lg}(1 + \cos\theta) \tag{4}$$

where γ_{sg} and γ_{lg} are the surface tension on solid-gas and liquid-gas boundaries, respectively; θ is the boundary wetting angle; solid is PU; liquid is EG.

The mean value of γ_{sg} was calculated as average of 5 different measurements and error of measurements did not exceed the value of 0,5 mN/m.

The spectra of luminescence were obtained using the luminescent spectrometer SDL-1 (LOMO) in an excitation by the mercury lamp. The emission of the most intensive line with the maximum on 365 nm was selected with light filter UFS-2.

Two-electrode method measurements of conductivity at a direct current (dc) were conducted using a Hiresta UP high resistivity meter (Mitsubishi Chemicals, Japan). A dc voltage of 10 V was applied across the sample thickness. The samples were dried over night in an oven at 40°C under vacuum and then kept in dried environment, for the elimination of any moisture effects.

Dielectric relaxation analysis was performed using dielectric spectrometer on the base on alternating current bridge R5083. Complex dielectric permittivity, $\varepsilon^* = \varepsilon' - i\varepsilon''$, of disc-like specimens (diameter: 20 mm) sandwiched between gold-coated brass electrodes was measured over the frequency window from 102 to 105 Hz in the temperature interval from --40 to 120 °C. They have been analyzed from the traditional point of view (Pathmanatham & Johari, 1990; Pissis & Kanapitsas, 1996). Additional formalisms such as: complex admittance σ^*, electrical modules M^* and impedances Z', Z'' were used according to formulas.

$$\varepsilon' = C_1 / C_o, \mathrm{tg}\delta = \omega RC_1 \text{and } \varepsilon'' = \varepsilon'\cdot \mathrm{tg}\delta \tag{5}$$

$$\sigma^* = \sigma' + \sigma'', \sigma' = \omega\varepsilon'' , \sigma'' = \omega\varepsilon', \tag{6}$$

$$M^* = M' + M'', M' = \varepsilon'' / \left(\varepsilon'^2 + \varepsilon''^2 \right), M'' = \varepsilon' / \left(\varepsilon'^2 + \varepsilon''^2 \right) \tag{7}$$

$$Z' = M'' / (\omega C_o), Z'' = M' / (\omega C_o) \tag{8}$$

C_o and C_1 – are instrument and standard capacitor capacities, ω – cyclic frequency.

The electron spectra of the copper (2+) containing PU films and of copper (2+) chelate compounds solutions in dichloromethane (c = 10^{-2}M) in the ultra-violet and visible region were recorded using the spectrometer Specord UV-VIS.

The quasi-elastic neutron scattering (QENS) was recorded using the multi detector spectrometer "NURMEN" on the atomic reactor BBP-M (The institute of the nuclear research of the NAS of Ukraine). The self-diffusion of chloform used as low molecular probe liquid in swelled PU films was analyzed.

2. Heterogeneity of metal containing polyurethanes

2.1. Structural heterogeneity of PU according to X-ray data

Formation of a polymer matrix in the presence of metal chelate compounds favours creation of a new hierarchy in structural organization of the polymer as compared with metal free system. This effect is caused by complex formation between metal chelate compound and functional groups of the forming polymer (Ying, 2002; Kozak et al., 2000).

Figure 1 represents the WAXS and Figure 2 presents SAXS intensity profiles of metal-free PU and PU modified with metal β-diketonate. The asymmetric diffuse diffraction maxima (Figure 1) point on the amorphous structure of the metal-free and metal containing CPU and LPU. For the LPUs the short-range order parameter d (equation 1) is equal to 0.44 nm and don't depend on the metal chelate compound amount (table 1). For the CPU the Bragg's period (d) changes from 0.44 to 0.46 nm with increasing of the modifier amount from 0,5 to 5% wt

The PU's SAXS profiles are characterized by the presence of one amorphous maximum with q_m positions varying from 1,7 to 2,0 nm^{-1} (Figure 2). Such maximum points on the existence of changeover period of uniform electron density scattering elements and areas of uniform distribution of hard and flexible blocks in PU. The Bragg's period (D) falls from 3,7 to 3,1 nm with increasing of the modifiers amount from 0,5 to 5% wt. (table 1).

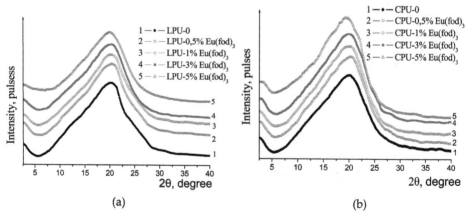

(a)   (b)

Figure 1. The WAXS intensity profiles of CPU (a) and LPU (b): metal-free (1), modified with 0,5% (2), 1% (3), 3% (4) и 5% (5) Eu(fod)₃.

Bottom-Up Nanostructured Segmented Polyurethanes with Immobilized in situ Transition and Rare-Earth
Metal Chelate Compounds – Polymer Topology – Structure and Properties Relationship

57

System	2θ, degree	d, nm	q_m, nm^{-1}	D, nm
CPU-0	20	0.44	1.7	3.7
CPU-0,5% Eu(fod)$_3$	20	0.44	1.7	3.7
CPU-1% Eu(fod)$_3$	19.9	0.45	1.76	3.6
CPU-3% Eu(fod)$_3$	19.9	0.45	1.76	3.6
CPU-5% Eu(fod)$_3$	19.4	0.46	2.0	3.1
LPU-0	20	0.44	1.7	3.7
LPU-0,5% EEu(fod)$_3$	20	0.44	1.7	3.7
LPU-1% Eu(fod)$_3$	20	0.44	1.9	3.3
LPU-3% Eu(fod)$_3$	20	0.44	1.8	3.5
LPU-5% Eu(fod)$_3$	20	0.44	1.9	3.4

2θ - the diffraction maximum angular position, degrees;
d – distance between PU atomic layers from WAXS, nm;
q_m - value at maximum intensity of $I(q)$ relationship, nm^{-1};
D - changeover period of uniform electronic density scattering elements from SAXS, nm.

Table 2. X-ray structural characteristic of LPU and CPU

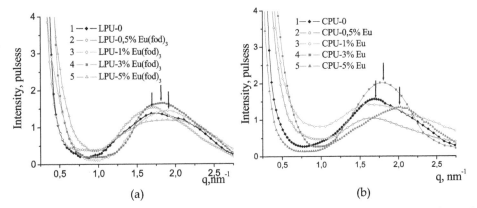

(a)   (b)

Figure 2. The SAXS intensity profiles of CPU (a) and LPU (b): metal-free (1), modified with 0,5% (2), 1% (3), 3% (4) and 5% (5) Eu(fod)$_3$.

Analysis according (Porod, 1982) of heterogeneity range (l_p) and average diameter (l_1, l_2) of different scattering elements in CPU-0, CPU-Cr and CPU-Co indicate existence of two types of nanosize heterogeneities in the bulk of PU. The first one (with $l_1 < D$) is inherent to segmented PU. The second one (with $l_2 > D$) is generated in the presence of transition metal chelate compound. We can define the latter structures as "metal chelate compound – polyurethane" complexes with polymer chains as macro ligands (Kozak et al., 2006; Nizelskii & Kozak, 2006) (Scheme 1).

Thus, the immobilization *in situ* of metal chelate compounds in polyurethane is accompanied with enrichment of polymer matrix with the nanosize heteroligand macro complexes of metal formed simultaneously with organic nanosize structures typical for metal-free polymer.

2.2. Dynamic heterogeneity of PU according to EPR data

The structural heterogeneity of PU influences the local segmental mobility of macro chains, resulting in "dynamic heterogeneity" of the systems. The analysis of mobility of SP introduced into the polymer gives information concerning such heterogeneity.

Calculated values of τ are listed in the table 2. They characterize the hindered rotation of SP in PUs of different topology. The greater value of τ is, the harder rotation of the probe occurs in polymer matrix.

System	$\tau \cdot 10^{-10}$, c	System	$\tau \cdot 10^{-10}$, c
CPU-0	45	LPU-0	48
CPU-1%Cu(eacac)2	43	LPU-1%Cu(eacac)2	69
CPU-1%Ni(acac)2	42	CPU-1%CuCd	45
CPU-1%Cr(acac)3	50	CPU-1%CuZn	32
CPU-1%Co(acac)3	49	CPU-1%CuNiCo	51

Table 3. The correlation time of TEMPO in CPUs and LPUs, modified with 1% of metal chelate compounds.

As it can be seen from the table 2, in CPU modified with 1%wt. Co(3+) and Cr(3+) chelate compounds the values of τ increase indicating reduction of SP mobility as compared with metal-free CPU. In the contrary, for CPUs modified with 1%wt. Cu(2+), Ni(2+) values of τ decrease as compared with metal-free CPU. This means that Co(3+) and Cr(3+) containing CPUs have more dense macro chain packing as compared with metal free CPU. Where as Cu(2+) and Ni(2+) containing CPUs possess looser macro chain packing. Similarly to (Lipatov et al., 2000), the effect we can relate to difference in metal chelate compounds electron configuration and symmetry. In addition, the influence of metal chelate compound on PU dynamic depends also on the polymer topology. For example, it can be seen the opposite influence of Cu(2+) chelate compounds on the macro chain mobility in LPU and CPU (Table2).

The analysis of SP EPR-spectrum shape and hyperfine splitting (HFS) gives additional information concerned probed medium. In PU matrices that contain metal chelate compounds the EPR spectra of SP have asymmetric shape (Figure 3). In all of the spectra occur essential increasing of central component and broadening of all components as compared with TEMPO spectrum in homogeneous glycerol matrix. In many spectra there is noticeable splitting of low-field and/or high-field components of SP spectrum.

The peculiarities observed are most likely the result of signal superposition of "fast" and "slow" probes located in polymer regions with different mobility. In conformity with above supposition the temperature increasing brings on enhancement of the SP EPR spectra isotropy (Figure 3). Initially asymmetric ESR spectrum becomes more isotropic while heating the sample. The spectrum components narrow and the intensity of central component diminishes.

Bottom-Up Nanostructured Segmented Polyurethanes with Immobilized in situ Transition and Rare-Earth
Metal Chelate Compounds – Polymer Topology – Structure and Properties Relationship

59

As a result of heating the equalizing of polymer segments mobility and „unfreezing" of „slow" SP rotation diffusion occurs. The correlation time decreases with the rise of temperature due to increasing of molecular mobility and "softening" of PU matrix. Figure 4 represents the relationship $\tau(T)$ for CPU.

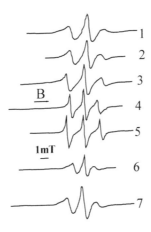

Figure 3. The spectra of the TEMPO introduced in CPU+%Er(acac)₃ at the various temperatures: 18 ºC (1); 26 ºC (2); 44 ºC (3) ; 90 ºC (4); 114 ºC (5); 21 ºC (30 min after thermal heating) (6); 18 ºC (2 days after thermal heating) (7).

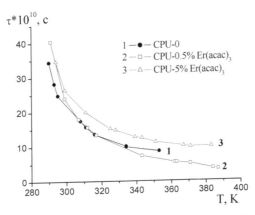

Figure 4. The thermal dependence of correlation time of the TEMPO in CPU-0 (1), CPU, modified with 0,5%wt. (2) and 5%wt. (3) of Er(acac)₃.

2.3. Thermodynamic heterogeneity of PU according to DSC data

The PU's thermodynamic heterogeneity is closely associated with above discussed types of heterogeneities. The influence of the metal chelate compounds on the thermodynamic heterogeneity and thermo-physic properties of PUs was analyzed by DSC. Figure 5 illustrates the temperature dependences of specific heat capacity of CPUs modified with 0,5; 1; 3; 5%wt. of Cu(acac)₂. The thermo-physic characteristics of copper-containing CPUs are given in Table 3.

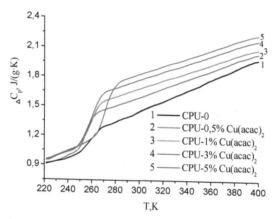

Figure 5. Temperature dependence of specific heat capacity for copper-containing CPU.

System	T_g, K	ΔT, K	ΔC_p, J/ (g·K)	$\frac{\Delta C_{p(CPU-Cu)}}{\Delta C_{p(CPU-0)}}$
CPU-0	258	18	0,25	1
CPU-0,5%Cu(acac)₂	256	16	0,38	1,52
CPU-1%Cu(acac)₂	258	18	0,43	1,72
CPU-3%Cu(acac)₂	260	21	0,50	2,00
CPU-5%Cu(acac)₂	271	20	0,55	2,20

Table 4. The thermo-physical properties of copper-containing CPU.

It is evident from fig. 5 and table 3 that for the CPUs the specific heat capacity (ΔCp) grows with increasing of Cu (2+) chelate content from 0,5 to 5% wt. comparing with CPU-0. In addition, the high temperature shifting of glass temperature (T_g) and the broadening of the temperature interval of glassing (ΔT) for CPU- 3%Cu and CPU-5%Cu are observed. The similar effect was discussed in (Lipatov et al., 1999) for CPUs, modified with 1%wt of various transition metals chelate compounds. That effect we can relate to formation of coordination bonds between functional groups of CPU and copper (2+) chelate compound.

Thus, growth of T_g and ΔC_p values with increasing of Cu(acac)₂ amount corresponds to rise of polymer segments with decreased mobility due to complexing.

The ratio of $\Delta C_{p(CPU-Cu)}$ to $\Delta C_{p(CPU-0)}$ allows estimate the degree of PU's thermodynamic heterogeneity (Bershtein. & Yegorov, 1990) and analyze the influence of metal chelate modifier content on this type of heterogeneity (table 3). As it can be seen the thermodynamic heterogeneity degree of CPU correlates with modifier amount in the system. This result agrees with X-ray data (section 2.1).

2.4. The formation of ordered micro regions in metal containing PUs

The segregation of metal containing micro crystals in CPU-5% Co and CPU-5%Cr was revealed in (Kozak et al., 2006):. Such unexpected segregation seemed unlikely due to homogeneous dispersion of metal chelate compound solution in reaction mixture (see 2.1) and coordination immobilization of metal chelate compounds in PU matrix. Nevertheless, the further X-ray study of CPU-5%Cu, LPU-5%Cu (fig. 6) and microscopy data (see 2.5) confirm partial segregation of metal-containing sites in PU matrices. This effect can be explained by different complex ability of segmented PU soft and hard components towards metal chelate compound as well as by higher mobility of PU's soft component.

The Scherer's equation (Stompel & Kercha, 2008) for the average diameter (L) of crystallite in amorphous media allows estimate dimensions of the particles in metal-containing PU.

$$L = k\lambda \ / \ (\beta cos\theta_m) \qquad (9)$$

Here X-ray wavelength λ = 1,54 Å, k is the shape factor assigned to 0,9, L is the average diameter of the crystals in angstroms, θ_m is the Bragg's angle in degrees, and β is the half-height of diffraction angle in radians. The value of L is equal to 3 nm for CPU-1%Co, it is equal to 4 nm for CPU-1%Cr and it is equal to 10 nm in LPU-1%Eu. The evaluated dimensions of the aggregates in copper containing PUs are ranged from 8 to 12 nm.

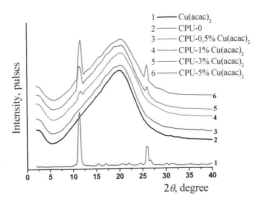

Figure 6. The WAXS diffractograms of the CPU-%Cu(acac)₂ films.

Segregation of the micro crystals detected via WAXS study has been also fixed by optical light transmission microscopy and by the scanning electron microscopy (SEM) (Figure 7). The micro crystals detected by optical microscopy are coloured like metal chelate compounds used as PU modifier. Such colouring indicates enrichment of the crystals with corresponding metal ions. The crystalline regions can be formed by the modifier itself and/or by complexes of modifier with PU chains as macro ligand. The last conclusion agrees with the X-ray data that register several discrete peaks in Cu(2+), Cr(3+) and Co(3+) containing PUs (Figure 6).

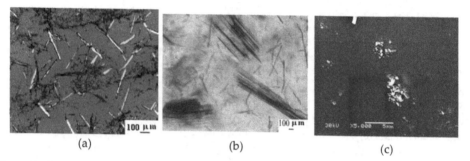

(a)   (b)   (c)

Figure 7. The optical microscopy (a, b) and SEM microscopy micro images of the LPU-0,5%Cu(acac)₂ (in polarized light) (a), LPU-5%Cu(acac)₂ (b) and CPU-5%Cr(acac)₃.

Optical microscopy allows obtain information concerning two surfaces of one PU film. One of them formed on the boundary "polymer-support" (the PU's surface formed on the Teflon support) and another formed on the boundary "polymer-air" (the PU's surface formed on the air).

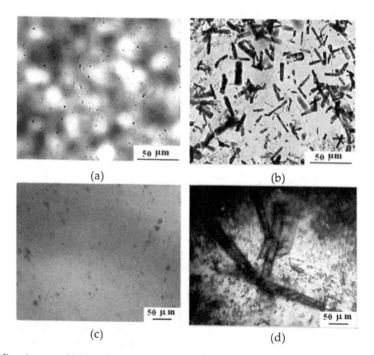

Figure 8. Micro images of LPU-1%Eu(fod)₃ (a,b) and CPU-1%Cr(acac)₃ (c, d) surfaces formed at the "polymer-air" boundary (a, c) and the "polymer- support" boundary (b, d).

Figure 8 illustrates the typical differences in surfaces of PU films. As it can be seen, at surface formed at the boundary "polymer-support" (fig. 8, *a*) the size and quantity of crystals are larger. Where as, at surface formed at the "polymer-air" boundary (fig. 8, *b*) the size and quantity of crystals are significantly smaller. For example, the mean size of crystals in LPU-1%Eu changes from the one surface to another from 20 μm to 0,5 μm.

Detailed analysis of PUs surface properties depending on the boundary nature can give additional information.

3. Influence of metal chelate modifiers on surface properties of polyurethanes

The presence of metal chelate compounds in reaction mixture can influence the surface tension of the formed polyurethane. In (Lipatov, 1997) the surface properties were studied of PU with metal ions introduced through in four different ways. There are filling, metal ion cross-linking, metal ion chain-extending and diffusion of metal chelate compound from its solution to polymer being formed earlier. It has been shown that the surface properties of metal containing PU depend on metal quantity much less than on the way of metal chelate compound introduction in polymer. For example, the γ_{sg} of PU filled with Cr(acac)₃ (0.18% wt.) changes up to 8 mN/m. On the contrary the γ_{sg} of Pb (15% wt) cross-linked PU changes up to 0.3 mN/m as compared with metal free PU.

Obviously, the PU's surface structure depends on the boundary "polymer-support" or "polymer-air". Data of ESCA and IR-spectroscopy by (Lipatova et al., 1987; Lipatova & Alexeeva, 1988) point on possibility of the chemical unequivalence of the polymer surfaces formed at the different boundaries. In addition in (Kozak et al., 2010) it was observed substantial difference in luminescence intensity at different surfaces of the PU films modified with europium (3+) chelate compounds. Therefore, the surface properties of europium containing LPU and CPU were compared for surfaces formed at the "polymer-air" and "polymer-support" boundary using measurement of contact wetting angle. The data obtained are listed in the table 4.

The values of surface tension of metal containing PU obtained using Wilgelmy method(with water as wetting liquid) (Lipatov et al., 1997) are consistent with values of the surface tension calculated using measurement of contact wetting angle (Table 5) of standard liquid.

The wetting angles at the „polymer-air" boundary for all of CPU and LPU are from 5.5 to 15.5 degrees less than the wetting angles at the „polymer-support" boundary (table 4). The difference between relative values of surface tension (γ_1-γ_2) takes values from 2.18 to 5.59 mN/m. As it is known, the higher compound polarity is the greater surface energy and surface tension it possesses. Obtained results allow conclude that PU surface formed at the „polymer-air" boundary is enriched with more polar groups (e.g. urethane) and PU surface formed at the „polymer-support" boundary is enriched with less polar groups (e.g. glycol segments).

Concentration of PU less polar groups that form the weak complexes with metal chelate compound at the „polymer-support" boundary can facilitate the partial segregation of metal containing centres at this boundary. That conclusion is consistent with microscopic data and photoluminescence measurements.

System	θ, degree		$\gamma_{EG\text{-}PU}$, mN/m		$\Delta\theta = \theta_2 - \theta_1$, degree	$\Delta\gamma = \gamma_1 - \gamma_2$, mN/m
	θ_1 (the „polymer-air" boundary)	θ_2 (the „polymer-support" boundary)	γ_1 (the „polymer-air" boundary)	γ_2 (the „polymer-support" boundary)		
CPU-0	55	65	38,06	34,50	10	3,56
CPU-1% Eu	53	64	38,74	34,67	11	4,07
CPU-3% Eu	51	66	39,39	33,95	15	5,44
CPU-5% Eu	56	64	37,70	34,77	8	2,93
LPU-0	58	70	37,00	32,45	12	4,55
LPU-1% Eu	61	67,5	35,95	33,43	6,5	2,52
LPU-3% Eu	67,5	73	33,43	31,25	5,5	2,18
LPU-5% Eu	59	74,5	36,63*	30,64	15,5	5,99

θ_1, θ_2 – the wetting angles at the „polymer-air" and the „polymer-support" boundaries, respectively, degree;
γ_1, γ_2 – the surface at the „polymer-air" and the „polymer-support" boundaries, respectively, mN/m;
* the unbalanced wetting angles

Table 5. The contact wetting angle (θ) and surface tension (γ) of PU films. The standard liquid is ethylene glycol (EG) $\gamma_{EG\text{-}air}$ = 48,36 mN/m.

Varying of the metal containing modifier amount (from 0,5 to 5% wt.) in CPU practically does not affect surface tension. In the contrary, change of metal chelate compound content in LPU from 0,5 to 3%wt. lead to decreasing of both γ_1 and γ_2.

The difference of the tendency in changing of surface tension in LPU and CPU clearly depend on polymer topology. Different PU topology results in different segmental mobility of the polymer, that agrees with DRS data. This effect described detailed in Section 5 and Section 2.4. At that time we can't formulate the certain reason for non monotonous influence of the modifier's amount on the surface tension.

4. The influence of polymer topology and modifier content on the luminescent properties of segmented polyurethanes

According to (Lobko et al., 2010) the PU matrix can intensify the photoluminescence of europium chelate compounds introduced into polymer *in situ*. Taking into account that immobilization *in situ* of metal chelate compounds in polymer matrix can influence both structure and properties of the hybrid system (Nizelskii & Kozak, 2006; Nizelskii et al., 2005) the investigation of rare-earth metal compounds in polymeric environment is a way for creation of new optically active materials.

Bottom-Up Nanostructured Segmented Polyurethanes with Immobilized in situ Transition and Rare-Earth
Metal Chelate Compounds – Polymer Topology – Structure and Properties Relationship

65

LPUs and CPUs modified with Eu(3+) chelates when exposed in 365 nm UV-light demonstrate the intensive photoluminescence in red region. Figure 9 represents the luminescent spectra of LPU and CPU, modified with various amount of Eu(fod)₃.

The luminescent spectra of europium containing PU are diffuse, while luminescence spectrum of Eu(fod)₃ is enough well-resolved. According to (Poluectov et al., 1989) the luminescence spectra of europium β-diketonate solutions contain bands corresponding to the $^5D_0$-^7F_i -transitions (where i = 0,1,2,3,4). The spectra of Eu β-diketonate in PU matrices demonstrate the intensive wide band of photoluminescence in the region of λ=610-635 nm ($^5D_0$-$^7F_2$-transition), narrow band λ=660 nm ($^5D_0$-$^7F_3$ –transition) and bands of $^5D_0$-$^7F_{0, 1, 4}$-transitions (580, 600,700 nm, accordingly) of low-intensity. It is possible to explain the diffuse spectrum of luminescence of europium containing PU in the region of λ =610-635 nm by distortion of the Eu (3+) chelate geometry in PU due to complex "polymer-metal chelate compound" formation and due macroligand steric hindrances.

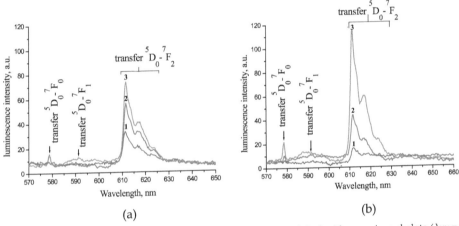

(a) (b)

Figure 9. The spectra of luminescence of LPU *(a)* and CPU *(b)*, modified with europium chelate (λ_{UV} = 365 nm): (1) 0.5%; (2) 1%; (3) 5%.

The intensity of PU-Eu luminescence depends both on the europium chelate content and polymer topology. The luminescence intensity increases with increasing of europium chelate compound content. The luminescence intensities of $^5D_0 \rightarrow {}^7F_2$ transition (λ=612nm) for LPU with 05%, 1% and 5%wt. of Eu(fod)₃, correspond as 1:1,8:2,4. The relationship of luminescence intensity *vs.* modifier percentage in CPUs is linear (1:3,3:9,2). The CPU-Eu with low modifier content has the lower luminescent intensity as compared with LPU-Eu. Where as CPU-5%Eu luminescence intensity is 1,5 higher, than LPU-5% Eu luminescence intensity. Taking into account data of Sections 2, 3, 6 we can suppose that due to difference in PU topology this effect is associated with higher concentration of polymer photo transmitting sites near the modifier in CPU as compared with LPU.

The tetra coordinated Eu (3+) chelate compounds with different additional ligands in an external coordination sphere were used to analyse the influence of additional coordination of europium chelate compounds on the intensity of their luminescence. The fig. 10, a illustrates the luminescent spectra of isolated Eu (3+) chelate compounds. The fig. 10, b represents the spectra of luminescence of CPU films, modified with 1%wt. of Eu (3+) compounds.

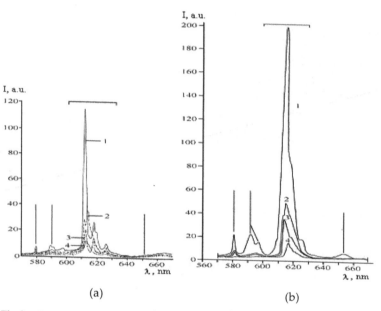

(a)   (b)

Figure 10. The luminescence spectra ($\lambda_{ex.}$ = 365 nm) of the europium (3+) chelate compounds (a) and of CPU films with 1%wt. of these chelate compounds (b): Eu(TTA)₃phen (1); Eu(TTA)₃TPPO (2); Eu(TTA)₃ (3); Eu(fod)₃ (4)

As it can be seen the luminescence of Eu (3+) chelate compounds introduced into PU matrix (only 1%wt.) is more intensive than luminescence of isolated metal chelate compounds (100% wt.). In addition, the intensity of luminescence of tetra coordinated Eu(3+) chelate compounds (Eu(TTA)₃phen and Eu(TTA)₃TPPO) both isolated and introduced into CPU matrix, considerably exceeds such intensity for 3-coordinated Eu(3+) chelate compound Eu(TTA)₃ that does not contain additional ligands The intensity of photoluminescence of Eu(fod)₃ also is considerably lower.

Estimation of Eu (3+) environment symmetry in various complexes via the coefficient of asymmetry (η) defined as ratio of intensity of $^5D_0 \rightarrow {}^7F_2$ –transition to intensity of $^5D_0 \rightarrow {}^7F_1$ transition. (Haopeng et al., 2008) shows that the greatest coefficient of asymmetry (η = 9) has CPU-1%Eu(TTA)₃phen characterized by the greatest intensity of luminescence. Consequently, the presence of additional ligand in the external coordination sphere of Eu (3+) favours increasing of luminescence intensity. Then increasing of Eu-chelate compounds luminescence intensity in PU can be explained in particular by additional coordination of lanthanide ion with the functional groups of PU and/or by formation of

Bottom-Up Nanostructured Segmented Polyurethanes with Immobilized in situ Transition and Rare-Earth
Metal Chelate Compounds – Polymer Topology – Structure and Properties Relationship

67

donor-acceptor complexes between aromatic fragments of PU and quasi-aromatic chelate rings of chelate compounds of rare-earth metals.

5. Dielectric relaxation and conductivity

The dielectric properties of PU were studied by broad band DRS measurements in wide range of temperature (-40 to 120 °C). The data are analyzed within the various formalisms. The direct current conductivity was both measured using two-electrode method and was estimated using DRS complex electric resistance $\sigma_{dc} = d/(AR_{dc})$ and $Z''(Z')$ isotherms (Cole-Cole diagram). Figure 11-13 illustrate obtained dielectric spectra. Calculated conductivity values are listed in Table 5.

According to two-electrode method the direct current conductivity of PU can drastically change in the presence of some metal chelate compounds. At the room temperature σ_{dc} for the CPU-5%Eu increases by one order as compared with CPU-0. In the presence of polyheteronuclear metal chelate compounds σ_{d} enlarges from 2 to 3 orders (fig. 13, table 5). DRS analysis of complex dielectric permittivity as well as complex admittance σ^*, complex electrical modulus M^* and impedances Z', Z'' allows reveal the nature of the observed conductivity.

System	a) σ_{dc}, Sm/cm 20°C	b) σ_{dc}, Sm/cm 40°C	System	a) σ_{dc}, Sm/cm 20°C	b) σ_{dc}, Sm/cm 20°C
CPU-0	1,78 10⁻¹²	1,3·10⁻¹¹	LPU-0	4,65·10⁻¹²	4.6·10⁻¹²
CPU-Cu	2,86 10⁻¹¹	2·10⁻⁹	LPU-Cu	4,25·10⁻¹¹	3.8·10⁻¹¹
CPU-Cu₂Zn	2,47 10⁻⁹	0,7·10⁻⁸	LPU-Cu₂Zn	1,51·10⁻⁹	1.2·10⁻¹⁰
*CPU²⁰⁰⁰-0	-	1*10⁻¹⁰	*CPU²⁰⁰⁰-Cu₂Zn	-	1*10⁻⁷
*CPU²⁰⁰⁰-Cu	-	1*10⁻⁹			

* The PU films synthesized with PPG-2000
a) σ_{dc} measured using two-electrode method and b) σ_{dc} obtained using DRS data

Table 6. PUs conductivity at a direct current

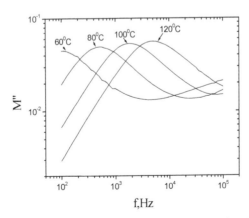

Figure 11. Log-log plots of the imaginary part of complex electrical modulus M'' *vs.* frequency for CPU–5%Eu at several temperatures

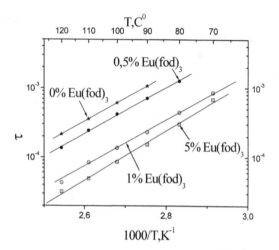

Figure 12. The thermal dependence of the relaxation time (τmax) for the CPU with various content of Eu (3+) chelate compound.

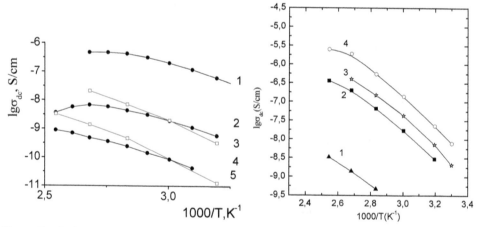

Figure 13. The lgσ_{dc} vs. 1/T for CPU: (a) CPU with various length of flexible component: CPU (PPG-2000) – 1%Cu$_2$Zn (1); CPU (PPG-1000) – 1% Cu(eacac)$_2$ (2); CPU (PPG-2000) – 1% Cu(eacac)$_2$ (3); CPU (PPG-2000) – 0 (4); CPU (PPG-1000) – 0 (5) and (b) CPU-0 (1) and CPU-Cu$_2$Zn formed in the presence of various solvents: 1, 4-dioxane (2); dichloromethane (3) and DMFA (4).

The curves on the fig. 11 have well defined maxima in temperature region of 60 to 120°C. According to (Pathmanatham & Johari, 1990; Kyritsis & Pissis, 1997) these maxima correspond to conductivity relaxation. Increasing of temperature is accompanied with shift of conductivity relaxation maxima to higher frequencies (fig. 11). The fact concerned to increasing of segmental mobility in PU. The metal chelate compounds introduction and increasing of their content in the system result in increasing of PU segmental mobility.

Experimental dependences of relaxation time ($\tau_{max} = \dfrac{1}{2\pi f_{max}}$) in log scale *vs.* 1/T (fig. 12) for metal-containing PU are linear indicating the Arrhenius-type of temperature dependence of τ_{max}.

Decreasing of τ_{max} value with the increasing of Eu(3+) content in PU confirm increasing of macro chains mobility in metal-containing CPU (Kozak et al., 2006). The activation energy of conductivity relaxation for CPU-0, CPU-0,5%Eu, CPU-1%, CPU-5% are approximately similar. Experimental dependences of $\log\sigma_{dc}$ vs. 1/T are non-Arrhenius both for PU-0 and metal containing PUs. It fit the theoretical curves of Vogel–Tamman-Fulcher (VTF) equation $\sigma_{dc}=\sigma_0\exp(-B/(T-T_0)$ (fig. 13) indicating influence of the PU free volume on charge transport. The results obtained give evidence of significant influence of structural organization in the modified PU on its conductivity level.

As it can be seen PU's direct current (σ_{dc}) conductivity grows with increasing of temperature and that is characteristic to ionic conductivity. The metal ion participation as current carrier is unlikely because to small amounts of metal ion in the modified PUs (~ 0.025-0.25% wt.). That fact and coordination immobilization of the modifiers in polymer makes unlikely increasing of the conductivity due to the metal chelate compound conductive properties. On the other hand ionic mechanism of conductivity and adequate amount of protons presented in PU matrix as well as observed increasing of polymer chain mobility in modified PU allows us to suppose proton participation in the process of charge transport.

Comparison of direct current conductivity of CPU based on PPG-2000 with conductivity of CPU based on PPG-1000 shows increasing of conductivity at the direct current of such system up to 10^{-7}Sm/sm at the 40°C due "softening" of PU. Nevertheless it can be seen that conductivity level of maximum soft LPU is at least one order lower then conductivity of CPU.

6. "Metal chelate compound - polymer" complexing and formation of the additional network of coordination bonds in metal containing PU

Mutual influence of metal chelate compound and polymer matrix due to complex formation is a decisive reason of observed changes of structural, dynamic, relaxation etc. characteristics of the metal contained PU. The complexing of metal chelate compounds with PU matrix was analysed using electron spectroscopy and EPR.

6.1. The complexing of the metal chelate compound with PU matrix according to the electron spectroscopy

The electron spectroscopy allows analyse both character of complexing of metal chelate compound with the polymer matrix and state of metal chelate compound in PU. The electron spectra of transition and rare-earth metal chelate compound in PU indicate presence of the band of *d-d*-transitions for the transition metal chelate compound introduced into PU (fig.14) and band of π-π-transitions for the rare-earth metal chelate compounds introduced into PU. That points on saving of chelate structure of the complexes in polymer matrix. While the change the band intensity, its broadening and shift to a long-wave region testifies their participation in complexing with PU.

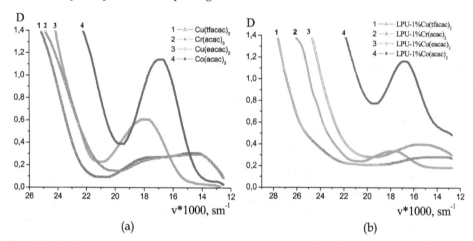

(a) (b)

Figure 14. The electron spectra of transition metal chelate compounds in dichloromethane (a) and LPU (b): Cu(tfacac)₂ (1), 1%Cr(acac)₃ (2), Cu(eacac)₂ (3); Co(acac)₃ (4); LPU Cu(tfacac)₂ (ε = 28 l/mol×sm) (1); LPU-1%Cr(acac)₃ (ε = 33 l/mol×sm) (2); LPU-1%Cu(eacac)₂ (ε = 40 л/ l/mol×sm) (3); LPU-1%Co(acac)₃(ε = 117 l/mol×sm) (4)

Figure 14 represents comparison of electron spectra in the visible region of the LPU films with 1%wt. of transition metal (copper, chrome, cobalt) chelate compounds (fig. 14, a) and the spectra of this metal chelate compounds dissolved (c= 10⁻² M) in dichloromethane (CH₂Cl₂) (fig. 14, b).

In addition to described above changes in electron spectra the influence of fluorine on complex ability of metal chelate compound in PU is evident due to, the rise of absorption level (ε = 40 l/mol×cm) and hypsohromic shift of maxima of band of d-d-transitions for LPU - 1% Cu(eacac)₂ as compared with LPU-1% Cu(tfacac)₂ that have fluorine in ligand (ε = 28 l/mol×cm). Fig. 15 illustrates the detailed analysis of electron transitions of copper ion in β-diketonates (4 transitions for D₂ₕ symmetry). Calculated maxima positions of Gaussian components of adsorption band corresponding to electron d-d-transitions of copper ion for Cu(tfacac)₂ and Cu(eacac)₂ in solution and in PU are listed in the table 6.

System	υ, cm⁻¹			
	$d_{чн} \to d_{z^2}$	$d_{чн} \to d_{x^2-y^2}$	$d_{чн} \to d_{xz}$	$d_{чн} \to d_{yz}$
Cu(tfacac)₂ in CH₂Cl₂	13658	15386	17930	19579
LPU -1% Cu(tfacac)₂	13435	15179	16322	17628
Cu(eacac)₂ in CH₂Cl₂	13715	15376	16686	18261
LPU-1% Cu(eacac)₂	12771	14497	16059	17738

Table 7. The allocation of Gaussian components of copper chelate compounds adsorption band.

Bottom-Up Nanostructured Segmented Polyurethanes with Immobilized in situ Transition and Rare-Earth
Metal Chelate Compounds – Polymer Topology – Structure and Properties Relationship

71

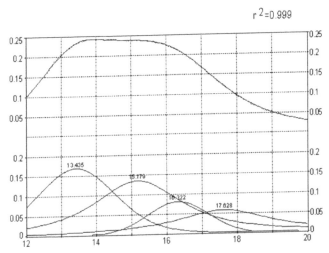

Figure 15. The adsorption band of LPU+1%Cu(tfacac)₂ with Gaussian components allocation.

The maxima of d_{xz} and d_{yz} transitions of Cu(2+) ion in copper chelate compounds, immobilized in PU matrix are shifted to the long-wave region, as compared to corresponding transitions of copper chelate compounds in solution (table 6). Visible broadening of the d_{z^2} component of absorption band of Cu (2+) chelate compounds in PU indicates the prevalence of axial coordination of macro ligand in "PU-metal chelate compound" complexes.

6.2. Complex formation in the "polymer-modifier" system according to EPR data

The EPR data confirm the complexing between the metal chelate compound and PU functional groups. The state of paramagnetic Cu(2+) containing chelate compounds in PU can be directly analyzed using EPR due to sensitivity of spin-electron parameters A_{II} and g_{II} (see table 7) of the tetragonal copper chelate compounds to symmetry and chemical nature of the copper nearest environment.

Decreasing of A_{II} and increasing g_{II} of Cu(acac)₂, Cu(tfacac)₂, Cu(eacac)₂ and (Cu₂Zn₂(NH₃)₂Br₂(HDea)₄)Br₂ immobilized in PU as compared with undisturbed compounds indicate the participation of the modifiers in the complexing with PU electron donor groups.

Figure 16 illustrates the representative EPR spectra of some copper containing modifiers both isolated and immobilized in PU with different topology (linear and cross-linked). The EPR spectra of polyheteronuclear powdered crystalline samples have anisotropic shape with weakly resolved HFS due to broadening of the spectrum components and possible tetrahedral distortion of the copper ion surrounding in the polyatomic complex. Immobilization of such chelate compounds in a PU network resulted in decreasing of the EPR signal intensity.

System	g_{II}	$A_{II} \times 10^{-4}$ cm^{-1}	$g_{\perp}$	$A_{\perp} \times 10^{-4}$ cm^{-1}	g_0	$a_0* 10^{-4}$ cm^{-1}
[1]Cu(tfacac)₂ *	2,271	187	2,052	23	-	-
CPU-1% Cu(tfacac)₂	2,275 / 2,290	172 / 151	2,059	21	2,131 / 2,136	71 / 64
LPU-1% Cu(tfacac)₂	2,283 / 2,302	162 / 141	2,049	19	2,127 / 2,133	67 / 60
[1]Cu(eacac)₂ *	2,276	187	2,055	22	2,128	68
CPU-1% Cu(eacac)₂	2,249	150	-	-	-	-
LPU-1% Cu(eacac)₂	2,298	173	-	-	-	-
[1]Cu(acac)₂ *	2,250	189	2,052	24	2,118	75
CPU-1% Cu(acac)₂	2,269	182	-	-	-	-
LPU-1% Cu(acac)₂	2,254	188	2,052	29	2,119	82
[2](Cu₂Zn₂(NH₃)₂Br₂(HDea)₄)Br₂	2,370	122	-	-	-	-
CPU-1% CuZn	2,370 / 2,300	132 / 142	-	-	-	-
LPU-1%CuZn	-	-	-	-	-	-

[1] undisturbed Cu(2+) complex in glassy matrix chloroform/toluene (40/60) (at -196°C)
[2] powder of polycrystalline sample
* (Lipatova & Nizelskii, 1972)

Table 8. Electron-spin parameters of isolated and polymer immobilized copper complexes.

The most reasonable explanation for this effect is distortion of the modifier's symmetry or geometry in PU-CuZn. The shape of EPR signal in PU network modified with (Cu₂Zn₂(NH₃)₂Br₂(HDe)₄)Br₂ indicates formation of complexes of various content and structure.

6.3. Low molecular probes dynamic and formation of additional network of the coordination bonds in metal containing polyurethanes

Using QENS and EPR with paramagnetic probes of various natures it was shown that complex formation of metal containing modifier with macro chains results in appearance of additional spatial obstacles for probe diffusion as compared with metal free network. The dynamic of low molecular probes and complex formation in the nanostructured polyurethane network containing Co (3+) chelate compounds immobilized *in situ* were analyzed.

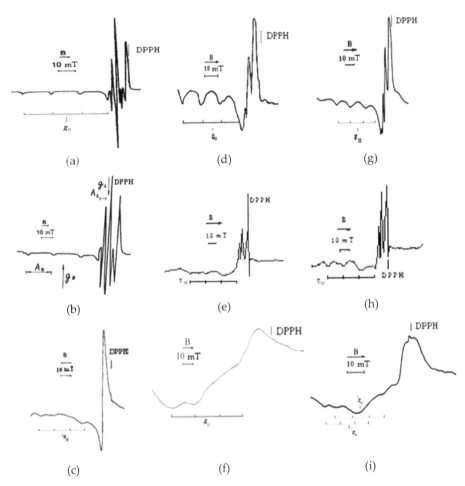

Figure 16. EPR – spectra of matrix isolated Cu(eacac)₂ (a), Cu(acac)₂ (b), (Cu₂Zn₂(NH₃)₂Br₂(HDea)₄)Br₂ (c) in chloroform-toluene at -196°C; LPU with 1%wt. of Cu(eacac)₂ (d), Cu(acac)₂ (e), (Cu₂Zn₂(NH₃)₂Br₂(HDea)₄)Br₂ (f) and CPU with 1%wt. of Cu(eacac)₂ (g), Cu(acac)₂ (h), (Cu₂Zn₂(NH₃)₂Br₂(HDea)₄)Br₂ (i).

According to EPR data obtained using complex spin probe (Kozak et al., 2006) it was demonstrated that in cobalt containing CPU the complexes "polymer-metal chelate compound" of two types are formed. The analysis of rotational diffusion of nitroxyl spin probe TEMPO (see table 2) reveals the decreasing of PU segmental mobility due metal chelate compound introduction and/or it content increasing. The dynamic of solvent molecules diffusion in swelled CPU-0, CPU-5%Co films and in probe liquid was analysed to compare the ratio of one-particle and collective modes of the solvent molecules motion.

System	$D \cdot 10^{-6}$, cm²/c	$D^F \cdot 10^{-6}$, cm²/c	$D^L \cdot 10^{-6}$, cm²/c	D^L/D, %
Probe liquid	3,52	3,18	0,35	8,0
The solution of Co(acaca)₃ 5%wt. in the probe liquid	3,33	3,04	0,29	8,7
CPU-0 (CH₂Cl₂)	2,71	2,40	0,31	11,5
CPU-Co5% (CH₂Cl₂)	1,85	1,32	0,53	28,6
CPU-Co5% (DMF)	1,25	0,89	0,36	28,8

where D^F – one-particle ("Frenkel ") diffusion coefficient ;
D^L – collective ("Lagrangian") diffusion coefficient.

Table 9. The diffusion parameters of the probe molecules in the swelled PU films

The sharp decreasing of both the general and one-particle component of diffusion coefficient for the metal containing PU as compared with the metal-free PU indicates the appearance of the spatial hindrances for the liquid molecule dynamics.

7. Conclusion

Immobilization *in situ* mono- and polyheteronuclear chelate compounds of transition and rare-earth metal in linear and cross-linked polyurethanes results in nanoscale structuring of forming polymer and is accompanied with polymer matrix enrichment by the nanosize heteroligand macro complexes of metal formed simultaneously with organic nanosize structures characteristic for metal-free polymer. Nanostructuring of formed in this way polyurethane favours creation of a new hierarchy in structural organization of the polymer as compared with metal free system as well as changes in dynamic, relaxation, optical, dielectric, surface etc. properties of the modified polyurethane.

Analysis of structural heterogeneity of metal-modified polymer indicates existence of two types of nanosize heterogeneities in the bulk of polyurethane. One of them is inherent to segmented PU and another is generated in the presence of transition metal chelate compound. The structural heterogeneity of PU influences the local segmental mobility of macro chains, resulting in "dynamic heterogeneity" as well as in "thermodynamic heterogeneity" of the systems.

The possible origin of the formation of the ordered micro regions is segmental structure of PU containing the soft and hard blocks with different complex ability relative to metal chelate compound. The PU's surface structure depends on the boundary "polymer-support" or "polymer-air". Concentration of PU less polar groups that form the weak complexes with metal chelate compound at the „polymer-support" boundary can facilitate the partial segregation of metal containing centres at this boundary.

The essential increasing of luminescence intensity of the rare-earth metal in the polyurethane environmental is a way for creation of new optically active materials. The intensity of PU-Eu luminescence depends both on the europium chelate compound content and polymer topology. Contrary to LPU the relationship of luminescence intensity *vs.* modifier percentage in CPUs is linear.

Increasing of the polyurethane conductivity to semi-conducting level is caused by the drastic increasing of macro chain mobility in the presence of polyheteronuclear modifiers. Conductivity level of LPU is at least one order lower then conductivity of CPU.

The results obtained indicate significant influence of structural organization of the modified polyurethane on its properties. The effect is caused by complex formation between metal chelate compound and functional groups of the forming polymer. The analysis of dynamic of low molecular probes and complex formation in the nanostructured polyurethane gives experimental evidence of existence of additional coordination bond network in metal-contained polyurethanes.

Author details

Nataly Kozak and Eugenia Lobko
Institute of Macromolecular Chemistry National Academy of Sciences of Ukraine, Ukraine

Acknowledgement

We would like to express our sincere gratitude to Professor Svetlana B. Meshkova (A. V. Bogatsky Physic-Chemical Institute of National Academy of Sciences of Ukraine, Odessa) and Professor Vladimir N. Kokozay (Taras Shevchenko Kyiv University) for synthesized heteroligand rare-earth metal's and polyheteronuclear chelate compounds, respectively.

8. References

Bershein, V.A. & Yegorov, V.M. (1990). *Differential Scanning Calorimetry in Polymer Physic.* (in Russ.). Chemistry, ISBN 5-7245-0555-X, St. Petersburg

Buchachenko, A. L., Wasserman, A.M., Alexandrova, T.A. et al. (1980). Spin Probe Studies in Polymer Solids, In: *Molecular Motions in Polymers by ESR*, R.F. Boyer & S.E. Keinath, (Eds.), 33-42, ISBN 3718600129, Harwood Academic Publisher, Chur, Switzerland

Bulavin, L.A.; Karmazina, T.V.; Klepko, V.V.; Slisenko, V.I. (2005). Neutron spectroscopy of condensed medium . (in Russ.). Akademperiodica, ISBN 966-360-009-8, Kyiv

Kovarskii, A.L. (1996). Spin Probes and Labels. A Quarter of a Century of Application to Polymer Studies, In: *Polymer Yearbook*, R.A. Pethrick, (Ed.), 113-139, ISBN 9783718659142, Harwood, Switzerland

Kozak, N.V.; Kosyanchuk, L.F.; Lipatov, Yu.S. et al. (2000). Effect of Zn2+, Cr(2+) and Ni2+ Ions of Cross-linked Segmented Polyurethanes. *Polymer Science*, Ser. A, Vol. 42, № 12, (June-August 2000), pp. 1304-1309, ISSN 0965-545X

Kozak, N.; Nizelskii, Y.; Mnikh, N. et al. (2006). Formation of Nanostructures in Multicomponent Systems Based on Organic Polymer and Coordination Metal Compound. *Macromolecular Symposia*, Vol. 243, (November 2006), pp. 243-262, ISSN 1521-3900

Kozak, N.V.; Lobko, Eu.V; Perepelitsina, L.M. et al. (2010). The Surface Characteristics of Polyurethanes, Contained β-diketonate Europium (3+). (in Ukr.). *Ukrainian Chemistry Journal*, Vol. 76, № 8, (June-August 2010), pp. 121-126, ISSN 0041-6045

Kratky, O.; Pilz, I.; Schmitz, P.J. (1966). Absolute intensity measurement of small-angle x-ray scattering by means of a standard sample. *Journal of Colloid Interface Science*, Vol. 21, №1, pp. 24-34, ISSN 1095-7103

Kyritsis, A. & Pissis, P. (1997). Dielectric studies of polymer-water interactions and water organization in PEO/water systems. *Journal of Polymer Science*: Ser. B, Polymer Physics, Vol. 35, (July 1997), pp. 1545-1560, ISSN 0965-545X

Lipatova, T. E. & Nizelskii, Yu.M. (1972). Complex Forming and Mechanism of the Catalysis of the Reaction Urethane Formation in situ β-diketonate of metals. (in Russ.). In: *The Successes of the Polyurethane Chemistry*, Yu. S. Lipatov, pp. 214-244, Naukova Dumka, Kyiv

Lipatova, T.E.; Shylow, V.V., Alexeeva, T.T.; et al. (1987), The Influence of the Support Character on the Structure of the Linear Polyurethanes Surface. (in Russ.). *Polymer Journal*, Seria B, Vol. 29, № 4, (December 1987), pp. 55-260, ISSN 1560- 0904

Lipatova, T.E. & Alexeeva, T.T. (1988). The spectroscopy investigated of the surface of segmented polyurethanes. (in Ukr.). *The Ukrainian Chemistry Journal*, Vol. 54, № 6, (June 1988) , pp. 324 – 628, ISSN 0041-6045

Lipatov, Yu. S.; Kosyanchuck, L. F.; Kozak, N. V et al. (1997). Effect of Metal Compounds on the Surface Properties of the Solid Polyurethanes being Formed in their presence. *Journal of Polymer Materials*, Vol. 14, № 3, (December 1997), pp. 63-268, ISSN 0970-0838

Lipatov, Yu. S.; Kozak, N. V.; Nizelskii, Yu. M. et al. (1999). Effect of Metal Compounds on the Physical Aging of Poly (urethanes). *Polymer Science*, Series A, Vol. 41, №8, (September 1999), pp. 1308-1315, ISSN 0965-545

Lobko, Eu.V; Kozak, N.V.; Meshkova, S.B. et al. (2010). The Photoluminescence of Polyurethanes, which Modified by the Chelate Compounds of threes- and tetra coordinated europium (3+). (in Ukr.). *Polymer Journal*, Vol. 32, № 5, (December 2010), pp. 410-415, ISSN 1818-1724

Nizelskii, Yu. N.; Shtompel, V.I.; Kozak, N.V. et al. (2005) The Nanostructure Heterogeneity of Polyurethane Films, Formed in the Presence of β-diketonate of Metals. *Report of National Academy of Science of Ukraine*, № 10, (October 2005), pp. 142-148, ISSN 1025-6415

Nizelskii, Yu. & Kozak N. (2006). In Situ Nanostructured Polyurethanes with Immobilized Transition Metal Coordination Complexes. *Journal of Macromolecular Science*, Part B, Vol. 46, (April 2007), pp. 97-110, ISSN 0022-2348

Pathmanatham, K. & Johari, G.P. (1990) Dielectric and Conductivity Relaxations in Poly (hema) and of Water in its Hydrogel. *Journal of Polymer Science*: B, Vol. 28., Issue 6, (May 1990), pp. 675-689, ISSN 0887-6266

Pissis, P.; Kanapitsas, A. (1996). Broadband dielectric relaxation spectroscopy at $10^{-4} - 10^{10}$ Hz. *Journal of the Serbian Chemical Society*, Vol. 61, № 9, pp.703-715, ISSN0352-5139

Poluectov, N.S.; Kononenko, L.I.; Yefrushyna, N.P. et al. (1989). Spectrometric and Luminescence Methods of lanthanides detection, (in Russ.), Naukova Dumka, ISBN 512000749X, Kyiv

Porod G. (1982). General theory. In: *Small-angle X-ray Scattering*. O.Glatter & O.Kratky (eds.)., 17-51, Academic Press, ISBN 0-12-286280-5, London

Saunders, J.H. & Frish, K.C., (1962). *Polyurethanes. Chemistry and Technology.* Part 1, Chemistry, Interscience Publisher, John Wiley&Sons, ISBN 0898745616, New York-London

Skopenko, V.V.; Garnovsky, A.D.; Kokozay, V.M.; et al. (1997). *The direct synthesis of coordination compounds*, (in Russ.), Ventury, ISBN 966-570-025-1, Kiev

Stompel, V.I. & Kercha, Yu.Yu. (2008) *The Structure of Linear Polyurethanes*, (in Ukr.), Naukova Dumka, ISBN 978-966-00-0699-7, Kyiv, Ukraine

Schmidt, P.W.; Hight, R.J. (1960). Slit height corrections in small angle x-ray scattering. *Acta Crystalographia*, Vol. 13, (June 1960), pp.480-483, ISSN 0365-110X

Tavana, H.; Gitiafroz, R.; Hair, M.L. et al. (2004). Determination of solid surface tension from contact angels: the role of shape and size of liquid molecules. *The Journal of Adhesion*, Vol. 80, (August 2004), pp. 705-725, ISSN 0021-8464

Vasserman, A.M.; Kovarskii, A.L. (1986). *Spin Labels and probes in physical chemistry of polymers"*, (in Russ.), Nauka, Moscow, Russia)

Vinogradova, E.A.; Vassilyeva, O.Yu.; Kokozay, V.N. (2002). An incomplete cube-like array of Cu, Zn, I with O atoms from 2-dimethylaminoethanol formed directly from zero-valent metal powders. *Inorganic Chemistry Communications*, Vol. 5, (May 2001), pp. 19-22, ISSN 1387-7003

Wirpsza, Z. (1993). *Polyurethanes: chemistry, technology and applications.* Elis Horwood Limited, ISBN 0-13-683186-9, London

Yang, H.; Zhang, D.; Shi, L. et al. (2008). Synthesis and strong red photoluminescence of europium oxide nanotubes and nanowires using carbon nanotubes as templates. *Acta materialia*, Vol. 56, (March 2008), pp. 955-967, ISSN 13596454

Ying, J. (Ed.) (2002). *Nanostructured Materials: Advances in Chemical Engineering,* Massachusetts Institute of Technology, ISBN 0-12-008527-5, Cambridge

Chapter 5

Polyurethane Flexible Foam Fire Behavior

Ahmadreza Gharehbagh and Zahed Ahmadi

Additional information is available at the end of the chapter

1. Introduction

Polyurethanes are a broad range of polymers, which are formed from the reaction between diisocyanates or polyisocyanates with diols or polyols. According to **the types and amounts of, polyols, isocyanate, ingredients** and the overall reaction circumstances, a broad range of products **like flexible foams, rigid foams**, elastomers, coatings and adhesives are produced.

Since the polyurethane products specially foams are playing an indispensable rule in our daily life **because of wide range of applications in automotive, household, refrigerators, insulations,** reducing of the fire risk of such a products are become more vital.

Conventional polyurethane **flexible** foams are easily ignited by a small flame source and burn rapidly with a high rate of heat release and smoke and toxic gases. This high flammability of polyurethane flexible foam is related to its cellular and open cell structure and low density of such foams. Oxygen can easily pass through the cells of the combustible material and in subjecting with an accident, smoldering cigarette or an electrical shock, foam catch fire [1].

Polyurethanes can be resisted against fire by different ways. **Depends on the types and applications of them, fire resisting could be done by the flame retardants using or by changing in** the polymer structure. In the whole picture the polymer ignition can be controlled by the following **factors.**

1. Extinguishing material reduction
2. Air supplying source reduction
3. Fire diffusion and heat generation reduction4. Increasing of the energy needed for entire combustion process

Different types of the fire retardants could be used according to one of the above mentioned categories. The flame retardants are acting according to one of the following mechanisms.

1. Reaction with the flame and preventing of the spreading of the fire by the created free radical blocks.
2. Preventing of the oxygen diffusion into the polymer
3. Lowering of the flame temperature with **removing energy from the system**
4. Char creation and creating a free place between the solid polymer and the disposed area.
5. Polymer expansion and making a free place between the fire and the decomposed polymer.

There are lots of materials which are known as the fire retardants according to the following groups.

- Halogenated flame retardants which are acting in the gas phase with disturbing the hydrogen-oxygen reaction. They react with hydrogen and create the halogen free radicals then they block free radicals of decomposed polymer.
- Metal oxides which act in solid or gas phase and some of the members of this group cause to reduce the flame temperature.
- Phosphorous containing compounds which create char on the extinguishing area of the polymer and prevent the oxygen feeding to the flame.
- Halogen free FR which the two main candidates are EG and Melamine. The heat stability of the polyurethanes especially the rigid foams at high temperatures depend on the isocyanurate to allophanates and biurets cross-linked bond ratio. Carbodiimide is produced by the condensation reaction of isocyanate with lose of CO2 (Fig.1). This reaction can be catalyzed by the cyclic phosphine-oxide. Generated carbodiimide is used as an unti-hydrolyze agent in the polyurethanes. The heat stability of the diverse products of polyurethanes is classified in Table (1).

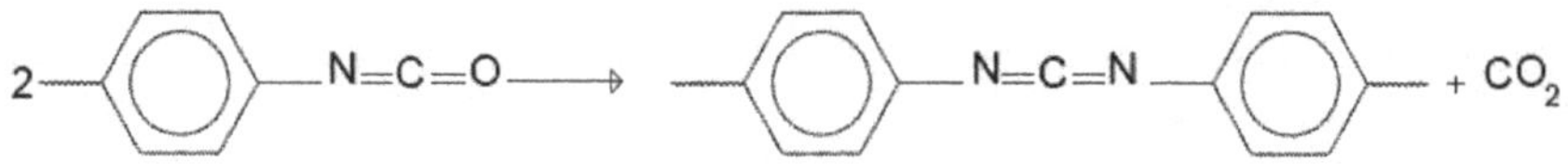

Figure 1. Condensation reaction of isocyanate to make Carbodiimide

Bonds	Decomposition Temperature(°C)
allophanate	100-200
biuret	115-125
Urea	160-200
Urethane	180-200
Substituted Urea	235-250
carbodiimide	250-280
Isocyanurate	270-300

Table 1. Heat stability of the diverse products of polyurethanes

2. Polyurethane foam morphology

Polyurethane morphology plays a vital rule on the fire properties of the polyurethane foam. The porous structure of the foam helps to diffuse the oxygen easily in to the foam and accelerate the ignition process. Fig.2 shows the SEM Picture of the polyurethane flexible foam with no filler inside. As it is clear, the cell structure of the foam includes Cell window, Strut and Strut joint [1].

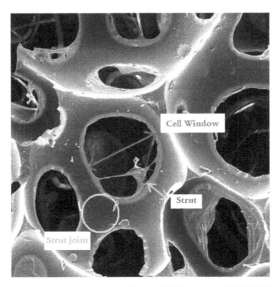

Figure 2. Fundamental concepts of polyurethane foam Cell Structure (SEM×200)

3. Polyurethane flexible foam fire retardants

3.1. Halogenated phosphorous flame retardants

In recent years the phasing out of some types of halogenated FR (flame retardant) due to persistence at environment and bioaccumulation and toxicity has been more investigated.

TMCP (Tris (2-chloroisopropyl) phosphate) and TDCP (Tris (1, 3-dichloroisopropyl phosphate) are two well-known liquid FR which are used in polyurethane flexible foam to make fire resisted **(Figure3-4)**. **Table (2) illustrate some important parameters of the mentioned fire retardants.** [2]

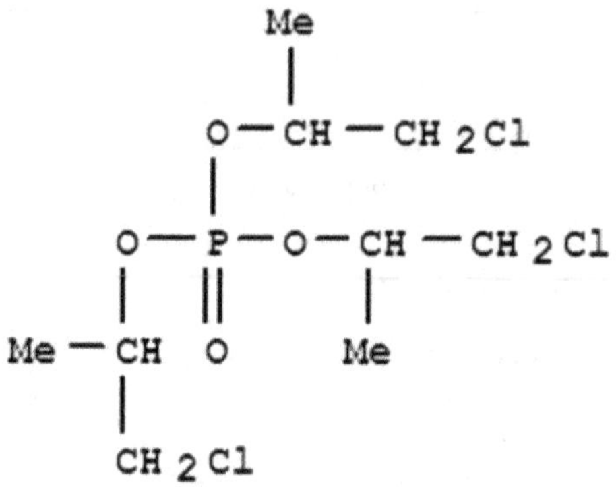

Figure 3. Chemical structure of TMCP

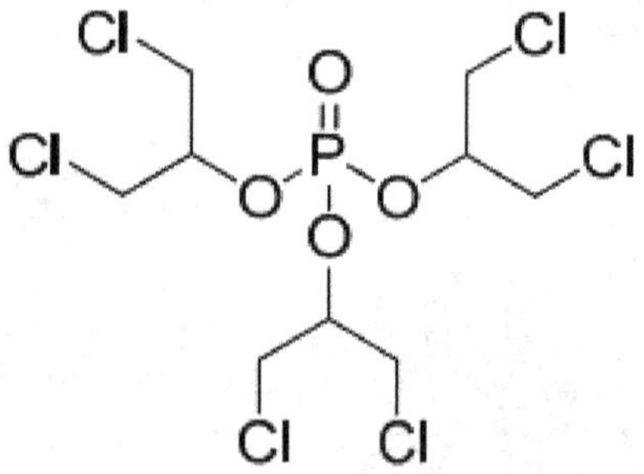

Figure 4. Chemical structure of TDCP

Property	TMCP	TDCP
CL content (%)	32.5	49
Mw(g/mol)	327.55	430.91
P content (%)	9.5	7.1
physical state at 25°C	clear liquid	clear liquid
Water solubility(%)	< 0.05	< 0.05

Table 2. Properties of TMCP and TDCP

Studies show that in the foams with only liquid FR (TMCP , TDCP) a very divergent combustion behaviour has been indicated. TMCP containing foams show lower TWL(total weight loss) and shorter burn time compared to TDCP containing foams.Moreover, TMCP containing foams didn't show any significant dripping and subsequent hole formation, a phenomenon seen at all levels of TDCP addition. TMCP and TDCP addition leads to decrease in the THE(total heat evolved) but an increase in the amount of smoke and carbon monoxide produced and this is why normally some amount of other FR such as melamine is added to the TMCP and TDCP containing foams to decrease total heat evolved, total smoke produced and CO emission significantly[2,3].

3.2. Halogen-free flame retardants

Due to the above mentioned reasons it has been a driving force to move toward the halogen free FR to compensate those weakness of halogenated FR, **despite of some disadvantages that the halogen free FR have e.g. they are mostly in solid state and they show process difficulties.** There are different types of halogen-free flame retardants which are behaving with different mechanisms. First group acts according to the expansion inside the polymer and oxygen-diffusion prevention and second group does by the cooling of the ignited surface of the polymer. One important example of the above mentioned groups are leaded by expandable graphite (EG) and Melamine powder respectively.

3.2.1. Expandable graphite

EG is a graphite intercalation compound in which some oxidants, such as sulfuric acid, potassium permanganate, etc. are inserted between the carbon layers of the graphite [4]. Fig.5 illustrates the chemical structure of Graphite, diamond and C60 [1].

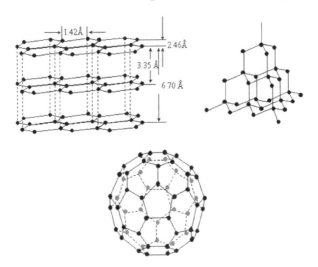

Figure 5. Comparison of Lattice graphite, Diamond and C60

When EG is subjected to a heat source, it expands to hundreds of times of its initial volume and creates voluminous, stable carbonaceous layer on the surface of the materials. This layer limits **the heat transfer from the heat source to the substrate and the mass transfer from the substrate to the heat source resulting in protection of the underlying material [5, 6]. The redox process [7]** between Sulfuric acid and graphite generates the blowing gases according to the reaction:

$$C + 2H_2SO_4 = CO_2\uparrow + 2H_2O\uparrow + 2SO_2\uparrow$$

The fire retardancy of EG is done by two steps [1]:

- The EG expands under the impact of Heat up to about 500 times of its original volume and creates a very large surface. It allows a quick oxidation of the carbon. The oxygen is taken out of the air and makes the air almost inert. This inert air extinguishes the fire.
- EG doesn't create flames while oxidation and will extinguish if no more heat will be applied to the glowing graphite. Therefore, no source of fire will be generated by the oxidizing graphite.

The more characteristic factors for EG which should be considered are:

- SET (start expansion temperature)
- Expansion volume
- Strength

Figure (6, 7) show particle size and distribution of two types of EG with different sizes (0.18mm, 0.25mm)

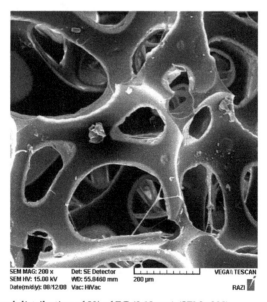

Figure 6. Particle size and distribution of 8% of EG (0.18mm) (SEM ×200)

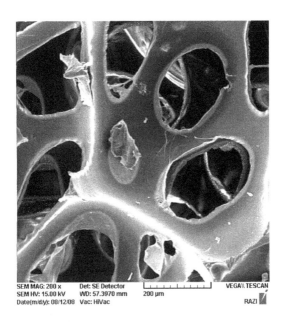

SEM MAG: 200 x   Det: SE Detector   VEGA\\ TESCAN
SEM HV: 15.00 kV   WD: 57.3970 mm   200 µm
Date(m/d/y): 08/12/08   Vac: HiVac   RAZI

Figure 7. Particle size and distribution of 8% of EG (0.25mm) (SEM×200)

3.2.2. Melamine

Melamine acts as fire retardant and smoke-suppressant according to the following combined mechanisms [8].

- Melamine is believed to act as a heat sink, increasing the heat capacity of the combustion system and lowering the surface temperature of the foam. Thus the rates of combustible gas evolution and burning are reduced.
- The nitrogen content of the melamine may partly end up as nitrogen gas when melamine burns, providing both a heat sink and inert diluents in the flame. The presence of melamine in the foam results in less heat generated by the flame, consequently less heat fed back to the foam and the rate of foam pyrolysis, i.e. generating of volatile fuel is reduced.
- Due to a chemical interaction between melamine and the evolved isocyanate fraction creating from degradation of polyurethane foam. This interaction reduces the amount of diisocyanate the main contributor to the smoke and CO release (Fig.8).

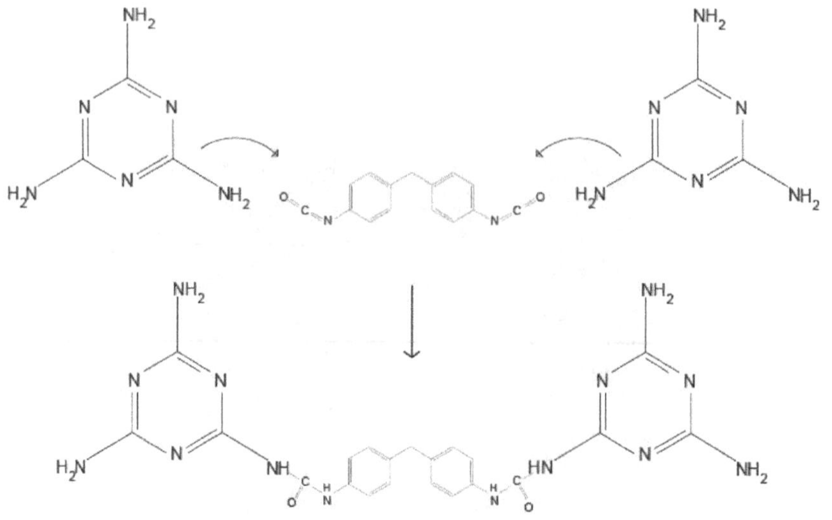

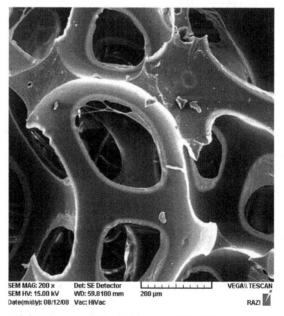

Figure 8. Chemical reaction between melamine and diisocyanate (MDI)

Figure (9) shows particle size and distribution of melamine powder inside the flexible foam.

Figure 9. Particle size and distribution of 8% of melamine (SEM ×200)

4. Properties of the polyurethane flexible foam with different types of fire retardants

Comparison between halogenated flame retardants which are mainly liquid with halogen free flame retardants (expandable graphite and melamine powder) which both are solid can be categorized as **four** items.

- Processing
- Reactivity
- Fire properties
- Physical properties.

4.1. Foaming process

The best choice for the processing as it is clear will be the liquid grade which has a good dispersion inside the polyol and less side effects.

Expandable Graphite has a limit pot life (3-4 hours) and when it subjects with the polyol component, it attacks to the catalyst of the polyol and destroys the catalyst during the foaming process. **For the foam producing, the highly recommendation is the EG containing polyol should react with proper isocyanate before the EG pot life reaches or the EG should be injected by an individual stream and mix with the polyol stream in the mixing head instead of pre-mixing with polyol.** Otherwise the produced foams will be collapsed. The other disadvantage of this technology is related to the fact that EG is very corrosive and make the mixing head to be damaged and it is preferred to be used a hardened grade of mixing head, a damaged mixing head needle picture is showed in (Fig.10)[9].

The advantages of this technology is the good homogeneity of the EG particles inside the polyol.

Figure 10. Mixing head needle corrosion by EG

Despite the EG, melamine has the longer pot life inside the polyol, which is around 24 hours but the fast sedimentation of the melamine powder in the polyol will be the main disadvantages so we need a suitable method to disperse the melamine powder in the polyol very well to achieve a homogeneous mixture.

4.2. Reactivity

Foam reactivity is determined by the following parameters:

-   Cream time (sec): Cream time is the time when the polyol and isocyanate mixture begins to change from the liquid state to a creamy and starts to expansion subsequently.
-   Gel time (sec): Gel time is the time the foam start to stiffen
-   Rise time (sec): rise time is the time the foam reach to its maximum height
-   Recession factor (%): the height percentage the foam is settled after 5 min after the rise time
-   Expansion factor (cm/kg): the proportion of the maximum height of the foam to foam weight.

The flame retardants would affect on the foam reactivity depend on the types of them, whether they are solid or liquid. Because they make changes in cell structure and total system heat capacity. The recession factor goes up with addition of EG and melamine but with different slopes. This is due to the increase in the average cell size of the foam. The bigger the flake size, the larger the cells and higher the recession factor. On the other hand melamine powders with small sizes are embedded on struts and joints and increase the viscosity and reduce the drainage rate which consequently, decreases the number of cells with bigger sizes [10]. Melamine powders with bigger size (bigger than struts and joints) are embedded inside the cell walls and open the cells.

Expansion factor which is related to the free rise density (FRD) reduces with addition of the EG and melamine in the foam. This is due to the increase the heat capacity of the entire system because of high heat capacity of melamine and EG. When melamine and EG content increases in the system, the heat capacity of the system increases and the system temperature reduces, therefore, the foam height and consequently expansion factor reduces [11].

4.3. Fire properties

The fire properties of the polyurethane flexible foams have been evaluated by different types of methods depends on the customer requirements. For example, the automotive, railway and airplane industries have their own standards.

The most important parameters which have been tested are: Cone calorimetry, flammability, smoke density and toxicity.

4.3.1. Cone calorimetry ISO 5660

The principle of the calorimetry by oxygen consumption (cone calorimeter) is based on the relation between the oxygen consumption and the heat release during the combustion. The ratio between the heat release and the weight of oxygen consumed is a constant (Huggett constant) equal to 13100 kJ/kg. It has been previously demonstrated that cone calorimeter results are in good correlation with results obtained in full scale fire test on upholstered furniture [3].

Samples of flexible foams (10*10*5cm) were exposed in a Stanton Redcroft Cone Calorimeter according to ASTM 1356-90 under a heat flux of 35kW/m2 (case of fully involved real firs). This flux was chosen because it corresponds to the evolved heat during a fire. An electrical spark igniter ignited volatile gases from the heated specimen. At least three specimens have been tested for each formulation. Data were recorded with a computer connected to the cone calorimeter. The test gives the opportunity to evaluate:

- RHR: Rate of Heat Release (kW/m2)
- Figra: fire growth rate: RHR/time (kW/m2/s)
- Weight loss (wt. %)
- Emission of carbon monoxide (ppm)
- TVSP: Total volume of smoke production (m3)
- THE: total heat evolved (kJ/cm2/g)

The combustion of flexible polyurethane foams is a two steps process (Fig.11).

The first step corresponds to the melting of the foam into a tar and the second to the combustion of the tar previously produced. [3]

These two degradation steps lead to two distinct peaks of rate of heat released.

- The RHR1 values (the values of RHR of the first and second RHR peaks).
- The T1 and T2 values (times at which RHR1 and RHR2 occur).
- The Figra2 values (the two maximum peaks on the Figra curve).

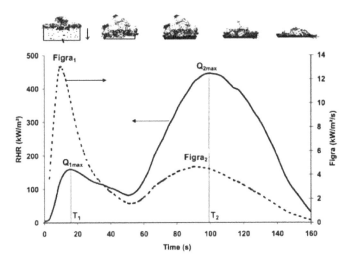

Figure 11. Combustion of flexible polyurethane foams: a two-stage process

4.3.2. Flammability

Flammability of the polyurethane foam is running with wide range of test methods depends on the applications and customers specification. Some fire tests standards include: FMVSS NO.302, British Standard 5852, ISO 9772 and FAA/JAA 25.853 Appendix F. as an example the airplane seat foam fire tests according to FAA/JAA 25.853 Appendix F have been investigated.

In this test 5 samples with 75mm*305mm*13mm dimension have been subjected with flame vertically for 12 sec and the following parameters have been investigated.

Burning time (the time that burning is continuing after removing the flame source) Burned length (the length of the foam which is damaged by the burning process) Time of dripping (the time which droplet continues to burn).

Synergetic effect

The synergetic effect of different types of FR has been observed. for instance, the fire properties of the EG loaded foams is much worse than when it is used by mixing with a liquid FR such halogenated phosphorous flame retardant. **Also when some amount of melamine is added to the TMCP and TDCP containing foams it helps to decrease total heat evolved, total smoke produced and CO emission significantly[2].**

Also the mixing of the liquid FR could boost the fire properties of the melamine loaded foam considerably.

4.3.3. Smoke density and toxicity

Smoke density and Toxicity are measured according to Airbus Directive ABD0031 (2005) on two categories:

1. Flaming
2. Non-flaming

Samples with 76mm*76mm*13mm are chosen to do the above mention tests against them in flaming and non-flaming status.

4.4. Physical properties

Physical and mechanical properties of the flexible polyurethane foams are evaluated by different types of tests in order to make an entire picture from the foam part performance during the consuming by the customer. For instance, flexible polyurethane foam is widely used as car seat foam and it has to keep its shape and other properties such as hardness and compression set during the time which is used. The most important properties of the polyurethane flexible foam as car seat foam according to RENAULT specifications are viewing as below.

- Core Density

- Compression Load Deflection (CLD: P25/5) and Sag-Factor according to D411003
- Compression Set according to D451046
- Tensile strength & Elongation at break according to
- Tear Strength according to D411048
- Resilience in 1st and 5th cycle according to D455128

When the polyurethane flexible foam is going to be fire resisted, some fire retardants in liquid or solid forms are entered in to the foam structure and make some changes in the physical properties of the final foam part. Mostly the valuable changes have been observed by the solid FR addition rather than the liquid one.

Depending on the fire retardant nature, shape and size, their addition may have some positive or negative effect on foam physical-mechanical properties. By loading the solid FR with the same amount, the foams become softer, because both additives have a similar size as cell windows and make the foam inhomogeneous. With EG, the homogeneity would be less than the foam loaded by melamine, because of its bigger size and flake shape which makes the foam much softer [1].

Sag-factor or the comfort index [12] increasing when the percentage of EG and melamine increases. It means that by adding solid FR, the comfort index would change considerably. Compression set, which is another very important factor, has increased by rising the EG percentage, but there was almost no changes in CS by increasing the melamine content. This effect is due to destroying effect of the cells structure by both additives but mainly by the EG.

Tear strensgth of the foams has improved by increasing the EG which could be related to the rigidity of EG flakes but deteriorates when melamine is added.

Finally, the resilience in 1st cycle is decreased for all additives but it is recovered in 5th cycle, because in 1st cycle the polymer chains have lost their flexibility due to rigid particles but after 5 cyclic movements the particles are embedded in struts and joints and the foam restores its flexibility.

5. Statistical method

Principle component analysis (PCA) is a useful method to illustrate relations between different parameters by using STAT-BOX-ITCF [13, 14].

Interpretation of the results consists first in the checking the representation of the variables in the circles of correlation. The correlations between variables are deduced from the relative position and the length of their corresponding vectors on the circle of correlation. An example of interpretation is done in (Fig.12); the angle between two vectors defines the intensity of the correlation (vectors 1 and 5). If α is=90^0, no relation exists between the variables. The strength of the correlation is higher when the angle is close to 0^0 or 180^0. So, orthogonal vectors (vectors 1 and 2) mean no correlation between the variables. Data are strongly correlated if their vectors are collinear (vectors 1 and 3, and vectors 1 and 4). The nature of the correlation also depends on the direction of the vectors: if vectors have the same direction (vectors 1 and 4) the variables are correlated, i.e. an increase in the variable linked to the vector 1 corresponds to an increase in the variable linked to the vector 4. Inversely, if vectors are opposite (vectors 1 and 3), the variables are anti-correlated.

The correlation between two variables is also a function of the length of the vectors. As example, vectors 2 and 6 are co-linear and so should be anti-correlated. But the weak length of the vector 6 means that its corresponding variable does not influence the variable linked to vector 2 [2].

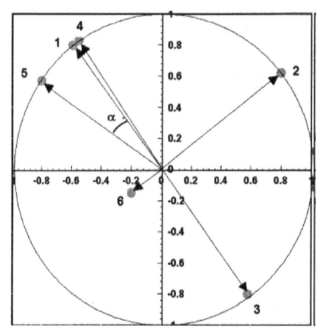

Figure 12. Interpretation of principal components analysis

5.1. Cone calorimeter–FMVSS 302

The principal components analysis from cone calorimeter and FMVSS 302 data shows the following correlations (Fig.13)

- *RHR1* is moderately correlated with *Figra1*: Figra1 is a variable that depends on the first peak of HRR (also called q1max), *d* the time it occurs. So, it seems quite coherent to find this kind of relation if the relative variation of the time is low.
- *RHR1* is correlated with *Figra2*. In the cone calorimeter, the foam degradation occurs in two main steps. It is obvious that an important consumption of fuel in the first step leads to a lower Figra2.
- FMVSS is strongly correlated with Figra1 and Figra2 and inversely correlated with *RHR2*. The lower Figra1 and Figra2, the slower the flame spread. A high RHR2 means loss of heat by dripping.

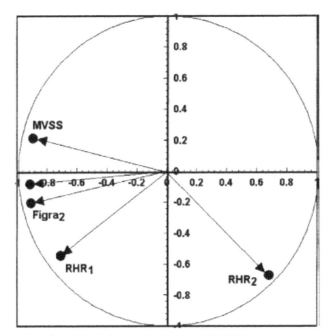

Figure 13. Correlation circle—relationship: cone calorimeter/FMVSS.

From the energy assessment of the foam consumption during 1s, we can find a relation between the propagation speed of the flame and the energy of the tar produced by the combustion (Fig.14).

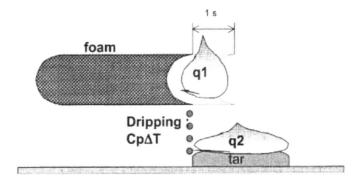

Figure 14. Principle of FMVSS.

As a first hypothesis, we can consider the following relation:

$$q1 + q2 - \Delta Q = Q = \text{constant}$$

- ΔQ corresponds to the part of heat used to melt the polymeric matrix leading to dripping.

- $q1$ corresponds to the energy released during the first stage of the combustion that leads to the formation of the tar (Figra1).
- $q2$ corresponds to the energy released by the combustion of the tar (Figra2).

This relation indicates the different strategies to decrease the value of RHR1 (and so Figra1), that is to say the flame spread in the FMVSS 302 tests:

- To decrease the total heat evolved Q using specific FR additives.
- To decrease the heat released during the first stage of degradation of the foam and as a consequence to decrease the heat fed back to the virgin polymer (decrease in Figra1).
- To increase RHR2, that is to say to reduce the energy of combustion by dripping.
- To delay the heat released by the tar. When the foam is molten, the tar starts to burn and this tar is not immediately lost by dripping. Hence, it is of interest to delay the combustion of this tar to enable it to drip (decrease in Figra2). An increase in RHR2 is not sufficient to reduce the flame spread and it is important that the high energy tar degrades at a later stage.

Hence, we may propose that the flame propagation rate in FMVSS 302 testing is much lower when easy melting and dripping allows heat reduction and tar dripping. It may be proposed that $q2$ corresponds in fact to the almost complete combustion of the tar.

Comparing the RHR curves of foams processed with variable water level, we note that the density of the foam strongly influences the first RHR peak (Fig.15). The higher the water content (the lower the density) the faster the step of melting under cone calorimeter conditions.

The effect of density on RHR1 may explain the previous correlation found between density and FMVSS 302. Low density leads to rapid melting and to a high flame propagation rate.

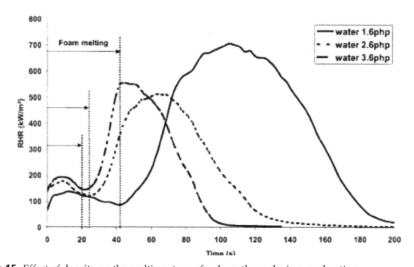

Figure 15. Effect of density on the melting stage of polyurethane during combustion

5.2. Cone calorimeter–British Standard

Ignition Source Crib 5 test to SI 1324 Sch. 1 Pt. 1

The statistical computation was made considering the two different sets of foams: the foams containing TMCP–melamine and the ones containing TDCP–melamine. The level of fire retardant additives has been included in the computation but is not shown on the circles of correlation.

Considering the TMCP–melamine foams (Fig.16) it is of interest to note that the lower are Figra1 and Figra2, the lower are the burn times, TWL and DWL. We also note that T2 is strongly inversely correlated with the data of SI 1324 Sch. 1 Pt. 1, that is to say the higher T2 the

Better results under the SI 1324 test (lower TWL, DWL and burn time).

The statistical computation of the data from the formulations TDCP–melamine clearly shows that the fire behavior of these foams in the SI 1324 test is linked to the second stage of degradation of the foam in the cone calorimeter (Figra2 and T2). Indeed, the Figra curve represents the fire growth rate of foam during combustion.

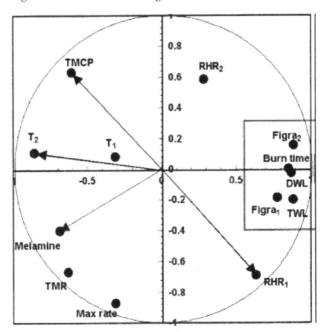

Figure 16. Correlation circle—relationship: cone calorimeter/SI

1324, TMCP–melamine formulations

A high Figra means a high rate of flame propagation and so leads to a high weight loss of the material. Hence, it is not surprising that Figra curves are strongly linked to the BS5852 results. The combustion of PU foam occurs in two steps: the "melting" of the foam and the combustion of the tar. The tar combustion is the most exothermic part of the combustion. A decrease in the heat released by the tar reduces the flame propagation and leads to a decrease in the weight loss of the foam (Fig. 17).

The TDCP and TMCP additives differ in their chlorine and phosphorus content and also in their temperature of degradation. TMCP degrades earlier than TDCP (150 °C and 210 °C, respectively); this temperature corresponds to a 5 wt. % weight loss under thermo gravimetric analysis conditions). A previous study [15] has clearly shown that TMCP is efficient in the early stage of combustion but no interaction with melamine is observed (temperature of 5 wt. % weight loss of melamine is 290 °C). TDCP acts later and when melamine starts to degrade about 50 wt. % of TDCP is available in the system, so a strong TDCP–melamine synergy is observed. The use of TDCP or TMCP in combination or not with melamine leads to very distinctive fire properties of the foams.

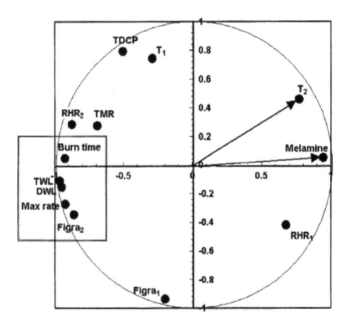

Figure 17. Correlation circle—relationship: cone calorimeter/SI

1324, TDCP–melamine formulations

Considering the TMCP–melamine foams, it is of interest to note that the higher the TMCP content the lower is RHR1. That confirms the early effect of TMCP that acts by decreasing the heat released by the foam in the first stage of the combustion. Secondly, the melamine content is inversely correlated with RHR2. As described previously, the temperature of decomposition of melamine is high (290 °C) and this inverse correlation indicates an efficiency of melamine during the combustion of the tar.

Regarding the TDCP–melamine formulations, we note a positive effect of the TDCP amount on the RHR1 peak. Even if TDCP degrades later than TMCP, a part of the TDCP is efficient in the first stage of the combustion.

The melamine content is strongly correlated with the SI 1324 data. High melamine content leads to a decrease in TWL, DWL burn time and maximum rate of weight loss.

The Figra2 and RHR2 peaks are also correlated with these data.

5.3. Properties–FMVSS 302

The statistical treatment shows that the FMVSS 302 rating is an inverse function of the density of the foam which is itself a function of the water index (Fig. 18). No significant relations may be proposed between FMVSS 302 and porosity or TDI index because data did not show any variation of the porosity (same SnOct content) and only a low variation of the TDI index.

The porosity index of the foam is strongly correlated with the SnOct range used in the foam manufacturing.

The previous study of conventional foams has revealed correlations between the FMVSS 302 testing and these parameters. The PCA study shows the absence of correlation between the EO content, the porosity (and so the SnOct range) and the index of the foam with the FMVSS 302 testing. However, it clearly shows that FMVSS 302 is strongly and inversely correlated with the density of the foam as it has been previously supposed.

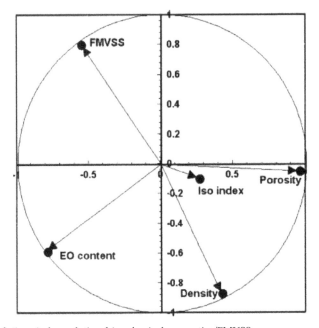

Figure 18. Correlation circle—relationship: physical properties/FMVSS.

Author details

Ahmadreza Gharehbagh
Iran Polyurethane Mfg.Co. NO.30, Tehran, Iran

Zahed Ahmadi
Color and Polymer Research Center, Amirkabir University of Technology, Tehran, Iran

6. References

[1] R. Bashirzadeh, A. Gharehbaghi, Journal of Cellular Plastics December 30, 2009, An Investigation on Reactivity, Mechanical and Fire Properties of Pu Flexible Foam

[2] Jerome Lefebvre and Michel Le Bras, Benoit Bastin and Rakesh Paleja, Rene Delobel, Journal fire sciences, Vol.21-september 2003

[3] Jerome Lefebvre, Benoit Bastin, Michel Le B, Sophie Duquesne,Christian Ritter, Rakesh Paleja, Franck Poutch , Flame spread of flexible polyurethane foam: comprehensive study, Polymer testing, 23(2004) 281-290

[4] Lei Shi, Zhong-Ming Li, Wei Yang, Ming-Bo yang, qiu-Ming Zhou, Rui Huang, Powder Technology 170(2006) P.178-184.

[5] Bourbigot, S.; Le, B.M.; Decressain, R.; Amoureux, J.P. J.Chem.Soc.-Faraday T.1996, 92(1), 149-158.

[6] Delobel R, Lebras M, Ouassou N, alistiqsa F. JFire Sci 1990; 8:85-108.

[7] Camino G,Duquesne S, Delobel R, Eling B,Lindsay C,Roels T.Fires and polymers. In: Nelson GL, Wilkie CA, editors.Materials and solutions for hazard prevention. Washington DC: ACS Pub; 2001.P.90.

[8] Dennis Price, Yan Liu, G.John Milnes, Richard Hull, Baljinder K.Kandola and A.Richard Horrocks Fire and Materials .2002; 26:201-206.

[9] Gharehbaghi, R. Bashirzadeh, and Z. Ahmadi, Polyurethane flexible foam fire resisting by melamine and expandable graphite: Industrial approach, Journal of Cellular Plastics, online published on 19 September 2011

[10] Turner, R.B., Nichols, J.B. and Kuklies, R.A. (1988). The Influence of viscosity in Cell opening of Flexible Molded Foams, In: Proceedings of the SPI, 31st Conference, Technomic, Lancaster, PA.

[11] A.Konig, U.Fehrenbacher and T. Hirth, E.Kroke, J.of Cellular plastics Vol.44-Nov.2008

[12] Kaneyoshi Ashida, Polyurethane and Related Foams, Chemistry and Technology (2007)

[13] G. Philippeau, Comment interpre'ter les re'sultants d'une analyse en composantes principales, I.T.C.F, Paris, 1986.

[14] H. Harman, Modern factor analysis, The University of

[15] Chicago Press, Chicago, 1976.

[16] B.Bastin, R. Paleja, J.Lefebvre, in: Polyurethanes EXPO 2002, API Conference, Salt Lake City, Utah,2002,p.244.

Chapter 6

Thermal Analysis of Polyurethane Dispersions Based on Different Polyols

Suzana M. Cakić, Ivan S. Ristić and Olivera Z. Ristić

Additional information is available at the end of the chapter

1. Introduction

1.1. Water-based polyurethane dispersions

Water-based polyurethane dispersions (PUD) are a rapidly growing segment of polyurethane (PU) coatings industry due to environmental legislations such as the clean air act and also due to technological advances that made them an effective substitute for the solvent-based analogs. Water-based or waterborne PUD have gained increasing importance in a range of applications, due in large part to properties such as adhesion to a range of substrates, resistance to chemicals, solvents and water, abrasion resistance and flexibility. Water-based PUD show very good mechanical and chemical properties and match the regulatory pressures for low volatile organic compound (VOC) containing raw paints. The continuous reduction in costs and the control of VOC emissions are increasing the use of water-based resins, motivating the development of PU dispersed in water. PU obtained from water-based PUD have superior properties when compared with similar materials obtained from organic media. Water-based PUD are used in many application areas to coat a wide range of substrates - for example footwear adhesives, wood lacquers for flooring and furniture, leather finishings, plastic coatings, printing inks and automotive base coats (Rothause et al., 1987; Kim et al., 1994; Ramesh et al., 1994).

Regarding the chemical nature of PU, the water based PU are applied with higher solids content, compared to the solvent based PU, because their viscosity does not depend on the molecular weight of PU, as is the case for solvent based PU (Gunduz & Kisakurek, 2004). Thus waterborne PUD can be prepared at high solid contents with a molecular weight enough to form films with excellent performance resulting solely upon "physical drying". This means that the film formation occurs by simple evaporation of water even at room temperature.

Waterborne PUD are fully-reacted PU systems produced as small discrete particles. 0.1 to 3.0 micron, dispersed in water to provide a product that is both chemically and colloidally stable, which only contains minor amounts of solvents and does not emit VOC. Polymeric structure of waterborne PUD is formed by usually reacting an excess of aliphatic isocyanates (mainly IPDI or HDI based), with a polyol or a mixture of polyols to form a prepolymer containing the so called soft segment. The polyols are generally polyesters, polyethers, or polycarbonates. The hard segment is generally formed by chain extending the prepolymer with short chain diamines and from the short chains containing ions. Due to incompatibility between the two segments of the polymer chain, the hard segment separates and aggregates into domains that act as reinforcing fillers to the soft segment. The degree of phase separation as well as the the the concentration of the hard segments are contributing factors to the good properties of PUD. PU backbone with a minority of the repeat units contains pendant acid or tertiary nitrogen groups, which are completely neutralized or quarternized, respectively, to form salts. Such ionomeric groups are absolutely necessary for the formation of dispersions, because they act as internal surfactants, and are not incorporated in the chain of the solvent-based PU. Ionic centers in the hard segment generally favor segregation and cohesion within the hard segment domains due to their strong electrostatic forces and thermodynamic incompatibility with the polymer matrix. Water-based PUD can be divided into two classes. The first group consists of polymers stabilized by external emulsifiers, and second one achieves stabilization by including hydrophilic centers in the polymer. Such hydrophilic centers may be one of the three types: non-ionic, cationic groups and anionic groups. These hydrophilic groups fulfill the function as internal emulsifiers and make possible to produce stable water/based emulsions. Water–based PUD can be classified into anionic, cationic and nonionic systems (Rothause & Nechtkamp, 1987; Kim et al., 1996).

Several processes have been developed for the synthesis of PUD. All of these have in common the first step, in which a medium molecular weight polymer (the prepolymer) is formed by the reaction of suitable diols or polyols (usually macrodiols such as polyether or polyester) with a molar excess of diisocyanates or polyisocyanates. In this reaction mixture, an internal emulsifier is added to allow the dispersion of the polymer in water; this emulsifier is usually a diol with an ionic group [carboxylate, sulfonate, or quaternary ammonium salt) or a nonionic group poly(ethylene oxide)]. The internal emulsifier becomes part of the main chain of the polymer. The critical step in which the various synthetic pathways differ is the dispersion of the prepolymer in water and the molecular weight build-up. The most important dispersions are emulsifier-free ionomer dispersions. The resulting dispersions are mainly anionic or non-ionic, that have the potential for wide variations in composition and property level. They can be obtained by different processes, however, the earliest process to prepare the aqueous PUD is known as acetone process and this process has remained technically important so far (Hepburn, 1992; Oertel, 1985). Within the last three decades several new processes have been developed such as prepolymer mixing process, hot melt process and ketamine/ketazine process.

The facts that aqueous/water PUD have become increasingly important for industrial and academic research in recent years is due to the following reasons:

1. the environmental law requires for the development of ecological-friendly products for which the emissions of volatile organic compounds (VOC) have been reduced to a minimum,
2. the economic reasons (substitution of expensive organic solvents in conventional PU with water),
3. the water PU surpasses performance of conventional isocyanate- and/or solvent-containing PU,
4. continuous increase in solvent prices, low raw material costs and easy to clean up the reactor system made waterborne PU system more popular in the industry.

2. Ingredients for water–based PUD

The basic components used to build up PUD include long–chain polyether, polyester or polycarbonate polyol, diisocyanate, aromatic or (cyclo)aliphatic, low molecular weight glycol and /or diamine, bis–hydroxycarboxylic acid and a neutralization base. In general, an excess of diisocyanate is treated with a long-chain linear polyol, bis-hydroxycarboxylic acid and other low-molecular-weight glycol to form an isocyanate-terminated prepolymer with a segmented structure. In this polymer, the long-chain polyol units form soft segments, and the urethane units-built up from diisocyanate, glycol and bis-hydroxycarboxylic acid form hard segments. The pendant carboxylic acid groups are neutralized with base to form internal salt group containing prepolymers that can be easily dispersed in water. The microphase separation between the incompatible soft- and hard-segment sequences contributes to the unique properties of PUD. The PU chains with NCO terminating groups can be extended with glycol forming urethane groups. Chain extenders are low molecular weight, hydroxyl and amine terminated. Aliphatic isocyanates: hexamethylene diisocyanate (HDI), isophorone diisocyanate (IPDI) and (4,4'–diisocyanatodicyclohexylmethane ($H_{12}MDI$), improve thermal and hydrolytic stability, resistance to UV degradation and they do not yellow (Bechara, 1998).

Aliphatic diisocyanates are less reactive than aromatic ones and they must be used with certain catalysts. Aromatic isocyanates: methylene diphenyl diisocyanate (MDI), toluene diisocyanate (TDI) and 1,5–naphthalenediisocyanate (NDI) on the other hand, provide for toughness but yellow upon exposure to UV light. Although early water dispersed PU resins heavily utilized TDI, there is a high tendency to shift to aliphatic diisocyanates or to the aromatic diisocyanates with NCO groups not directly attached to an aromatic nucleus (Gunduz & Kisakurek, 2004).

The two key classes of polyols are polyethers and polyesters. Polyester polyols have been largely used in PUD paints as they exhibit outstanding resistance to light and aging. There are four main classes of polyester polyols: linear or lightly branched aliphatic polyester polyols (mainly adipates) with terminal hydroxyl groups, low molecular weight aromatic polyesters for rigid foam applications, polycaprolactones, polycarbonate polyols. Polyether polyols are susceptible to light and oxygen when hot, however, they improve water dispersion, and impart chain flexibility. These are made by the addition of alkylene oxides, usually propylene oxide, onto alcohols or amines which are usually called starters or 'initiators'. Polyether based on propylene oxide contains predominantly secondary hydroxyl end–groups. Secondary hydroxyl end–groups are several times less reactive with isocyanates than primary hydroxyl groups and for some applications, polyether based only on propylene oxide may have inconveniently low reactivity. The primary hydroxyl content may be increased by a separate reaction of the polyoxypropylene polyols with ethylene oxide to form a block copolymer with an oxyethylene tip.

In the choice of polyol for PU application, selected polyols must be competitive with other polyols and also enable the final PU product to be cost competitive with other materials in the end application.

In a typical anionic PUD process, anionic groups (carboxylic and sulfonic) are introduced along the length of the polymer chain by using hydrophilic monomers or internal emulsifiers. As the hydrophilic monomer, dimethylol propionic acid (DMPA) is the most widely used acid, which has two hydroxyl groups, therefore, it can be one of the main constituents of the PU backbone. DMPA improves the hydrophilic property by serving as the potential ionic center with N–methyl pyrrolidone as the co–solvent. Tartaric acid (TA) can also be used, but it usually results in branching. However, TA improves the mechanical properties of PU paint. Study has shown that the particle size of dispersion depends on the content of DMPA. Therefore, increased amount of DMPA leads to more ionic centres in the PUD backbone and thereby increasing hydrophilicity of the polymer and hence reductions in particle size (Dieterich, 1981; Jacobs & Yu, 1993; Rosthauser & Nachkamp, 1986).

The chain extension step also has a high influence on the properties of the resin produced, not only due to the structure and concentration of the extender but also due to process variables that influence the particle size distribution. Chain extenders are difunctional glycols, diamines or hydroxyl amines. If a diol of low molecular weight reacts with the NCO terminated PU chains in the chain extension reaction step, urethane linkages will be formed but if a diamine is used as chain extender, the NCO terminated PU chains will form urea linkages. The higher density of hydrogen bonds of polyurea hard segments is responsible for the improved mechanical properties of polyurea and PU/urea products. Typical chain–extending agents are as follows: water, diethylene glycol, hydroquinone dihydroxyethyl ether, bisphenol A, bis(hydroxyethylether), ethanolamine, hydrazine, ethylene diamine. Aliphatic diamines such as hydrazine or ethylene diamine are used as chain extenders in processes directed to preparing waterborne PUD. In the chain extension step, it is most important to control the extremely fast reaction between NCO groups and NH_2 groups accompanied by the viscosity rise. Molecular weight of PUD increases by the formation of urea linkages through the chain extension step and it is the most important step to determine the molecular weight of PUD (Kim, 1996; Delpech & Coutinho, 2003).

In the synthesis of PUD, to neutralize the carboxyl and/or sulfo groups, are used agents that contain one or more bases and to form internal salt groups containing prepolymers that can be easily dispersed in water. During the neutralization, carboxyl and/or sulfo groups serve for anionic modification or stabilization of the PUD. Tertiary amines and in particular triethylamine are preferably used. The structure of waterborne anionic PUD is illustrated in Scheme 1.

Most commonly used catalysts in PU chemistry are tertiary amine catalysts and metal catalysts, especially tin catalysts. Tertiary amines are catalysts for the isocyanate–hydroxyl and the isocyanate–water reactions. Organotins are the most widely used, however organomercury and organolead catalysts are also used but have unfavourable hazardous properties.

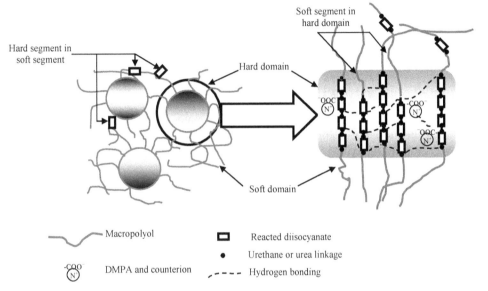

Scheme 1. Structure of waterborne anionic PUD (Tawa & Ito, 2006)

2.1. Various methods for preparing water–based PUD

The most important process is the prepolymer mixing process that has the advantage of avoiding the use of a large amount of organic solvent. In this process hydrophilically (carboxylate molecule) modified prepolymer is directly mixed with water. If the mixture viscosity is too high, a small amount of a solvent such as N–methyl pyrrolidone can be added before the dispersion step. Chain extension is accomplished by the addition of di– or polyamines to the water–based prepolymer dispersion.

The acetone process can be considered the link between the solvent synthesis and the prepolymer mixing process. In effect, the prepolymer is synthesized in a hydrophilic organic solvent, for example acetone solution and afterwards it is subsequently mixed with water.

The hot melt process explains the process of obtaining a PUD by the reaction of NCO–terminated ionic modified prepolymer with, for example ammonia or urea resulting in a prepolymer with terminal urea or biruet groups, respectively. The terminal urea or biruet prepolymer is methylolated with formaldehyde and mixed with water, forming dispersion spontaneously. By polycondensation (lowering the pH, increasing the temperature), chain–extension or cross–linking was obtained.

Ketamine and ketazine process explains the process of obtaining a PUD by reaction of NCO–prepolymers containing ionic groups mixing with a blocked amine (ketamine) or hydrazine (ketazine) without premature chain extension. These mixtures can be emulsified with water even in absence of co–solvents. The reaction with water liberates the diamine or hydrazine, which then reacts with the prepolymer.

Non–ionic dispersions are obtained similar to ionomer dispersions if the ionic centre is replaced by lateral or terminal hydrophilic ether chain. The temperature of dispersing process has to be kept below 60 °C. Non–ionic dispersions are stable towards freezing, pH changes and addition of electrolytes.

3. Thermal analysis of PU

Thermal analysis techniques have been used for many years in many scientific and industrial laboratories for studying the thermal decomposition of polymeric materials. Among them thermogravimetry (TG) is one of the most common since the mass of a sample is easy to measure accurately and valuable information regarding the nature of the process can be extracted from a mass loss against time or temperature plot. The thermal decomposition of PU (their degradation attributed to absorbed thermal energy) is important phenomenon from both fundamental and industrial applications (Pielichowski et al., 2004). The understanding of degradation processes allows determination of optimum conditions for designing PU in order to obtain high-performance polymer materials. Fundamental research has established that the thermal decomposition of PU is a complex heterogeneous process and consists of several partial decomposition reactions (Scaiano, 1989). The study of the decomposition of PU is particularly difficult since they degrade with the formation of various gaseous products and a number of decomposition steps are typically observed in thermogravimetric analysis (TGA) experiments. Some authors claim that the study of the thermo-degradation behavior of PU at high temperatures provides a fingerprint of the material that has to do not only with the characteristics of the original material, but also with its processing and the final quality of the end use products (Prime et al., 1988). The thermal stability of a material is defined by the specific temperature or temperature-time limit within which the material can be used without excessive loss of properties (Chattopadhyay & Webster, 2009). With respect to commercial applications, the investigation of thermal decomposition processes has two important aspects. The first concerns the stabilization of a polymer to obtain novel materials with a desired level of thermal stability that will be able to fulfill the demands of contemporary materials engineering. The second is to understand material behavior at higher temperature as well as to obtain characteristic thermal decomposition data.

A waterborne PUD can be defined as a binary colloidal system in which PU particles are dispersed in a continuous aqueous medium. PUD are usually prepared as low molecular weight NCO-terminated prepolymers for ease of dispersion. Then, diamines are generally used to increase the molecular weight by reaction with the terminated NCO end groups (chain extension). The presence of ionic species in PUD has a considerable effect on the physical properties. PUD are now one of the most rapid developing and active branches of PU chemistry.

PU are synthesized by the prepolymer reaction of a diisocyanate and a polyol (mainly polyethers and polyesters). If a diol of low molecular weight reacts with –NCO-terminated prepolymers in the chain extension reaction step, urethane linkages will also be formed but if a diamine is used as chain extender, the reaction between the –NH_2 groups and the –NCO terminated prepolymers will form urea linkages. In this case, poly(urethane-urea)s, which are the most important class of polyureas, are produced. These copolymers show reduced plasticity in comparison to homopolyurethanes. The resulting PU or poly(urethane-urea) chains consist of alternating short sequences forming soft (flexible) and hard (rigid) segments. The soft segments, originated from the polyol, impart elastomeric characteristics to the polymer. The hard segments are mainly produced by reacting the isocyanate and the chain extender. They are polar and impart mechanical properties to PU. The hard segments contain the highly polar urethane linkages. Due primarily to interurethane and urea hydrogen bonding, the two segment types tend to phase-separate in the bulk, forming microdomains. The hard segments act as physical crosslinks and, as a consequence, the physical, mechanical and adhesive properties depend strongly on the degree of phase separation between hard and soft segments and interconnectivity of the hard domains. The urethane linkages in PU can serve as H-bond acceptor and donor. In polyether-based PU, the urethane –NH can bond to either the polyether –O– linkage or the urethane –C=O groups. In the case of poly(urethane-urea) formation, there is an additional –NH from urea linkage participating in the interactions (Delpech & Coutinho, 2000). The degradation of thermoplastic PU has been extensively studied, and a number of reviews are available (Lu et al., 2002; Fambri et al., 2000). Thermal degradation of ester- and ether-based thermoplastic PU is performed under vacuum, air and nitrogen, allowing investigators to determine the mode of degradation (Dulog & Storck, 1996).

Polyester-based thermoplastic PU exibit rapid degradation in air and nitrogen, indicating that a nonoxidative mechanism is involved. In contrast, the significantly improved thermal stability of ether-based PU under vacuum and nitrogen indicates that the oxidative process plays a major role in the decomposition of ether-based thermoplastic PU. In general, the ester-based PU normally exibit better thermal and oxidative stabilities than the ether-based ones. The mechanism of thermal degradation of PU is very complex due to the variety of products formed.

It is proposed that the thermal degradation of thermoplastic PU is primarily a depolycondensation process, which starts at about 200 °C (Cakić et al., 2006 a). The first stage of decomposition is because of degradation of hard segments and starts at about 200 °C and at ~ 360-380 °C, while the second step of degradation is because of degradation of soft segments and ends above 480 °C. Waterborne PU should exhibit some different features in thermal degradation due to their unique chain structure, for example, salt-forming groups. Therefore, it is necessary to analyses their thermal degradation behavior to understand the structure-property relationship. The properties of PUD are mainly determined by the interactions between the hard and soft segments, and by the interactions between the ionic groups. The ionic group content, solids content, segmented structure, molecular weight of the macroglycol, the type of chain extender and the hard/soft segments ratio, determine the properties of PUD.

In the following sections, we will review typical results to demonstrate the utility of TGA in deducing the structural and bonding information about waterborne PUD based on different polyols. The thermal stability of PU and poly(urethane –urea)s cast films with anionomer character, obtained from waterborne dispersions and based on isophorone diisocyanate (IPDI), dimethylolpropionic acid (DMPA), poly(propylene glycol) (PPG), polycarbonate diol (PCD) and glycolized products obtained from recycled poly(ethylene terephthalate)(PET) is also compared. Three types of chain extenders were used: ethylene glycol (EG), propylene glycol (PG) and ethylene diamine (EDA). The effect of type of polyols, chain extender, type of catalyst, ionic content, length of soft segment, hard segment content and the presence of urea or urethane linkages on the thermal stability of the waterborne anionic PUD are discussed.

4. Water-based PUD based on poly(propylene glycol) and selective catalyst

One of the inherent drawbacks of waterborne PU technology is the formation of carbon dioxide due to the side reactions of isocyanate with water. When an isocyanate reacts with water, the products are a urea linkage (via an amine intermediate) and carbon dioxide. The carbon dioxide formation is problematic in that it causes imperfections in the coating during cure, such as blistering and pin-hole formation. The main aspect in the development of waterborne PU is in the first place to find methods for preventing the undesired secondary reactions with water and achieving the best crosslinking. This reaction is reduced to a minimum by the use of non-tin catalysts. One novel approach to control the water side reaction is the use of catalysts which selectively catalyze the isocyanate-polyol reaction and not the isocyanate-water reaction (Colling et al., 2002; Blank &Tramontano, 1996).

The relative selectivity (S) obtained from equation $S = P_{urethane}/P_{urea}$, was measured as urethane IR peak area ($P_{urethane}$)/ urea IR peak area (P_{urea}) ratio, by method given by Blank (Blank et al., 1999). After the integration of characteristic absorption max of urethane (1700 cm^{-1}, 1540 cm^{-1}) and urea (1640 cm^{-1}, 1570 cm^{-1}) was done, the relative selectivity was calculated. The manganese catalyst, a complex of Mn(III)–diacetylacetonatomaleate with various ligands based on acetylacetonate and maleic acid, used in some of the experiments (Stamenković et al., 2003; Cakić et al., 2006), has shown a high selectivity for the isocyanate–hydroxyl reaction in comparison to the commercially available zirconium catalyst (Blank et al., 1999). Zirconium catalyst is a proprietary zirconium tetra-dionato complex in the reactive solvent with the metal content of 0.4%.

TG is a suitable method to evaluate the thermal properties of several types of PU elastomers. The thermal stability of PU has been studied extensively because of the great importance of this group of materials (Chang et al., 1995). These thermoplastic elastomers generally are not very thermally stable, especially above their softening temperatures (Wang & Hsieh, 1997), and their mechanism of thermal degradation is very complex due to the variety of products formed. Commonly, it presents a bimodal profile where the first mode is related to the hard segments of PU. Usually, at a low heating rate, the degradation process results in differential weight loss (DTG) curves with several peaks, which is an indication of the complexity of the degradation (Delpech & Coutinho, 2000).

4.1. Experimental

Poly(propylene glycol) (PPG)), (M_n = 1000, hydroxyl value 111 mg KOH/g, dried under vaccuum, at 120 °C), isophorone diisocyanate (IPDI) and dimethylolpropionic acid (DMPA), were obtained from Aldrich Chemical Co. 1-methyl-2-pyrrolidone (NMP), dimethyl formamide (DMF) and triethyl amine (TEA) were received from Merck (Darmstadt, Germany). Ethylene glycol (EG) and propylene glycol (PG) obtained from Zorka Co.(Šabac Serbia). Dibutyltin dilaurate (DBTDL), was supplied by Bayer AG. Zirconium catalyst (ZrCat) was supplied by King Industries Inc., Norwalk, CT, USA. Manganese catalyst (MnCat) has been used in the reactive diluent with a metal content of 0.4% (Stamenković et al., 2003; Cakić et al., 2006 b). Water–based PUD from PPG with selective catalyst were prepared using the prepolymer method has been described in detail in our previously work (Cakić et al., 2009).

In the first step, PPG and DMPA were dispersed in DMF to obtain a homogeneous mixture and heated at 70 °C. IPDI and DBTDL were added to the homogenized mixture at 80 °C. An NCO/OH equivalent ratio of 3.0 was used. Hard : Soft segment ratio was defined as a ratio between IPDI weight and polyol weight in the starting formulation and is calculated as 1.5. The reaction times were determined by the dibutyl amine back titration method. After obtaining completely NCO terminated prepolymer, the mixture was cooled to 60 °C and the carboxylic groups were neutralized with TEA (DMPA equiv) dissolved in NMP. In second step the chain extension was carried out with EG or PG. The selective catalysts, ZrCat or MnCat, at concentration of 2% relative to the resin solids, have been added to the reaction solution. Water was added to the mixture and stirred to obtain dispersion of organic phase in water. The waterborne PUD contains 40 wt% solids (Cakić et al., 2009). Films were prepared by casting the waterborne dispersions on leveled surfaces and allowing them to dry at room temperature, for 7 days, and then at 60 °C, for 12h (Coutinho, 1996, 2003). When chain extenders were EG and PG, the PUD had to be cast in Teflon surfaces due to the high adhesiveness observed on the glass surface, making demoulding impossible. After demoulding, the films were kept into a desiccator to avoid moisture.and polyurethanes were formed.

4.2. Results and discussion

Degradation profile of waterborne PUD is influenced by the variation of chain extender presented in Fig.1. It was verified that the thermal stability was influenced by chain extender type. In a general way, thermal stability was higher when EG chain extender was used, in comparison to PG, probably because of the higher reactivity of the primary hydroxyl groups. The onset temperatures calculated for the first stage for PU chain-extended with EG were about 234 °C and 140 °C for PG. Unsymmetrical structure of IPDI enables easier diffusion of EG (Gunduz & Kisakurek, 2004).

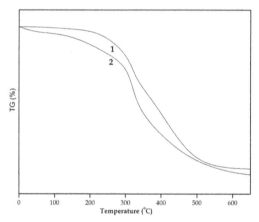

Figure 1. TG curves of PUD without catalyst with EG (1) and PG (2) as chain extender

Fig. 2 shows the degradation profile of PUD with variation of catalyst using catalysts of different selectivity. EG and PG formed urethane linkages by reaction with terminal NCO groups. The initial onsets observed are: 234 °C when EG was employed, 275 °C and 311 °C, when the chain extender was EG with ZrCat and EG with MnCat, respectively. The initial onsets observed are: 140 °C when PG was employed, 290 °C and 305 °C, when the chain extender was PG with ZrCat and PG with MnCat, respectively. The thermal stability was

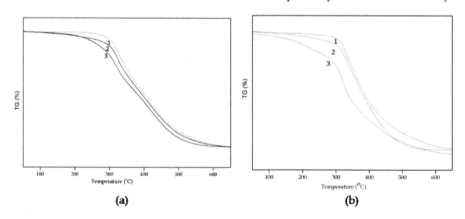

Figure 2. TG curves of PUD with EG as chain extender (a), PG as chain extender (b), and MnCat (1), ZrCat (2), without catalyst (3)

higher when MnCat was used, in comparison with the use of ZrCat. This result suggested that all the residual NCO groups in PU particles did not react with the chain extender completely. Because the viscosity of particle is high at low temperature in the chain extension step, it would take long time for chain extenders to diffuse into the particle. Therefore, the efficiency of chain extension increased as total surface area of particles increased (Jang et al., 2002; Cakić et al., 2007 a). In general, the presence of more selective catalyst has also been found to have a stabilizing effect on the resultant PU, as can be observed in the curves obtained for the samples chain-extended with EG and PG, in comparison to the samples obtained without selective catalyst probably due to favoring of isocyanate-polyol reaction and not the isocyanate-water reaction. The PUD was formed with higher hard segment proportions.

The DTG curves show that there are different stages of degradation which are not perceptible in TG curves, showing the close relation and mutual influence between the degradation process of hard and soft segments.

Fig. 3a shows the DTG curves corresponding to the TG degradation profiles presented in Fig. 2a, in which the catalysts were varied (MnCat or ZrCat). The chain extender employed was EG. Two peaks are observed. The first group of peaks, corresponding to the degradation of rigid segments formed by urethane and urea linkages, presents maximum of the peak from 200 to 250 ⁰C. The second group, related to the degradation of PPG soft segment, varying maximum of the peak from 375 to 418 ⁰C . The peaks shifting towards higher temperatures resulting from addition of more selective catalyst confirm the assumption that all isocyanate groups had not reacted with the added chain extender. Selective catalyst isocyanate-polyol reaction causes greater incorporation of chain extender in hard segments, which is reflected on higher thermal stability of hard segments (Jang et al., 2002; Lee et al., 1995; Cakić et al., 2007 b).

Fig. 3b depicts DTG curves related to TG profiles observed in Fig. 2b, in which the catalyst were varied (MnCat or ZrCat). The chain extender employed was PG. The first group of peaks presents maximum of the peak appearing in the range from 160 ⁰C to 195 ⁰C. The

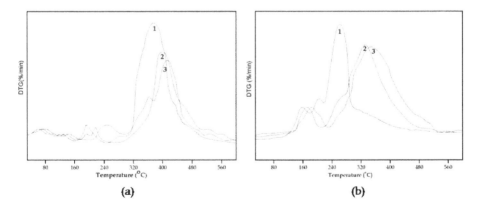

Figure 3. DTG curves of PUD with EG as chain extender (a), PG as chain extender (b), and without catalyst (1), ZrCat (2), (3) MnCat

second group can be observed in the range, varying maximum of the peak, from 267 to 347 °C, for PPG soft segments. A marked difference can be observed, promoted by changing the type of chain extender in DTG profiles, especially in the first stage of weight loss, corresponding just to urethane (EG or PG as chain extender) linkage degradation. The soft segment, formed only by PPG degradation step seemed to be also affected. The rigid segment formed from EG retarded the weight loss of PPG chains (peak at 375 °C), while PG showing peaks at 267 °C (Cakić et al., 2006, 2007 c). All DTG curves showed that there are different stages of degradation which are not perceptible in TG curves, showing the close relation and mutual influence between degradation of hard and soft segments.

The degradation profiles of PU cast films obtained from water-based dispersions were influenced by the type of chain extender, length of the hard segment and type of catalysts. The presence of more selective catalysts, which formed urethane linkages with higher hard segment proportions, had a marked influence on the degradation of the polymers, especially in elevated quantities, improving the thermal stability of the materials. The DTG curves showed that the length of the hard segment had a strong influence on the thermal profile of the samples as a whole. The type of chain extender, forming urethane linkages, affected the whole process of degradation and the presence of more selective catalyst improved the thermal resistance of the chains.

5. Water-based PUD based on glycolized products obtained from recycled poly (ethylene terephthalate)

Recycling of polymers has received a great deal of attention (Atta et al., 2006, 2007). Although several methods have been proposed for recycling waste poly(ethylene terephthalate) (PET), it is suggested that the most attractive method is glycolysis of chemicals into the corresponding monomers or raw chemicals that could be reused for the production of plastics or other advanced meterials (Patel et al., 2007).

Two-stage PUD synthesis was applied: the first, glycolysis of PET using different types of glycols (PG), triethylene glycol(TEG) and poly(ethylene glycol) (PEG 400), with different molar ratio of PET repeating unit to glycol (1:2 and 1:10); the second, preparation of PUD of the products formed.

PUD are prepared by anionic dispersion process (Cakić et al., 2011), using IPDI, glycolyzed products, DMPA as potential ionic center which allow water dispersibility and ethylene dimine (EDA) as chain extender.

5.1. Experimental

Example 1 of glycolysis reaction: Small pieces of PET waste (100 g), equivalent to 0.5 mol repeating unit (mol.wt. 192 gmol^{-1}) were added to 88.64 g PG (mol.wt. 76.09 gmol^{-1}), 173.07 g TEG (mol.wt.150 gmol^{-1}) or 461.5 g PEG 400 (mol.wt. 400 gmol^{-1}), such that the molar ratio of PET repeating unit to glycol was 1:2.

Example 2 of glycolysis reaction: In the second experimental runs of depolymerisation, appropriate amount of PET waste were added to 396.1 g PG, 750 g TEG or 2173 g PEG 400, so that molar ratio of PET repeating unit to glycol was 1:10. These mixtures (with different molar ratio PET/glycol) and 0.5 wt.% zinc acetate (based on the weight of PET as transesterification catalyst) were charged to a glass reactor, which was fitted with stirrer, reflux condenser, nitrogen inlet and temperature controller. This reactor was immersed in an oil bath and the content of the reaction kettle was heated at 190 °C for 2 h, subsequently the temperature was raised to 210 °C until all the solids disappeared.

The obtained glycolyzed oligoester polyols were analysed by the hydroxyl value (HV) determination according to the conventional acetic anhydride/pyridine method (Cakić et al., 2011). The hydroxyl value of the oligoester polyol obtained in the glycolysis reaction based on molar ratio of PET repeating unit to glycol, 1:2, with PG was HB_{PG}=490 mg KOH/g, TEG HB_{TEG}=370 mg KOH/g and PEG HB_{PEG400}=297 mg KOH/g.

The hydroxyl value of the oligoester polyol obtained in the glycolysis reaction based on molar ratio of PET repeating unit to glycol, 1:10, with PG was HB_{PG}=201 mg KOH/g, TEG HB_{TEG}=209 mg KOH/g and PEG HB_{PEG400}=192 mg KOH/g.

5.2. Synthesis of PUD based on glycolyzed products with molar ratio PET/glycol, 1:2

Anionic PUD based on glycolyzed products with molar ratio PET/glycol, 1:2, were prepared by modified acetone process. Acetone was added to the prepolymer and the dispersion is formed by the addition of water to this solution. Procedure for synthesis of anionic PUD has been developed adjusting the molar ratio of DMPA as a hydrophilic monomer to IPDI as 1:3.3. Mass of oligoester polyol, obtained by PET glycolysis, according to example 1, with a hydroxyl number which is equivalent to the hydroxyl number of 0.06 mol of poly(propylene glycol) PPG_{1000} (110 mg KOH/g), was for PG 15 g, TEG 20 g, poly (ethylene glycol) (PEG 400) 25 g.

The oligoester polyol and hydrophilic monomer (8 g, equ. 0.06 mol) was mixed in the cosolvent DMF (50:50 w/w), in a 250-ml round four-neck glass reactor connected to a stirrer, a thermometer, a reflux condenser and a nitrogen gas inlet. The reaction was carried out at 70 °C for 30 min to obtain a homogeneous mixture and the uniform distribution of hydrophilic monomer to PU backbone. IPDI (44.4 g, equ.0.2 mol) and catalyst DBTDL (0.03% of the total solid) were added to the homogenized mixture at 80 °C for about 4h until the amount of residual NCO groups reached a theoretical value, as determined by the dibutyl amine back-titration method. To reduce the viscosity and obtain a homogenous mixture of NCO prepolymers, acetone was added 50 wt% to the solid reaction mass. The theoretical value of NCO groups for PUD based on oligoester polyol obtained from the glycolysis with molar ratio of PET repeating unit to glycol, 1:2, was 19.2% for PG, 17.7%, for TEG and 16.1% for PEG.

After obtaining completely NCO terminated prepolymer, the mixture was cooled to 60 °C and the carboxylic groups in hydrophilic monomer were neutralized with TEA (DMPA equ). TEA was dissolved in NMP by stirring the solution for 60 min.

PU anionomer was cooled to 30 °C then dispersed in water (50% of total mass) under high speed stirring for 30 min. The rate of water addition to the mixture was carefully controlled, to obtain a stable inversion. Upon completing the phase inversion, EDA (0.03 mol) was added for 60 min. at 35 °C. PUD of about 30 wt% solids was obtained upon removal of acetone by rotary vacuum evaporation at 35 °C.

5.3. Synthesis of PUD based on glycolyzed products with molar ratio PET/glycol, 1:10

Anionic PUD based on glycolyzed products with molar ratio PET/glycol, 1:10, were prepared by prepolymer mixing method in two steps: synthesis of NCO-terminated prepolymers and the preparation of dispersions by introducing anionic centers to aid dispersions (Athawale & Kulkarni, 2009). The prepolymer mixing method has been developed adjusting the molar ratio of DMPA to IPDI as 1:3, which was increased compared to the previous procedure. Mass of oligoester polyol, obtained by PET glycolysis, according to example 2, with a hydroxyl number which is equivalent to the hydroxyl number of 0.1 mol of poly(propylene glycol) PPG_{1000} (110 mg KOH/g), was for PG 54.5 g, TEG 52.6 g, poly (ethylene glycol) (PEG 400) 57.3 g. The mass of oligoester polyols and hydrophilic monomer was the same as in the previous procedure, but the mass of IPDI was (66.6 g, equ.0.3 mol). determined by molar ratio of a hydrophilic monomer to IPDI as 1:3. In order to obtain NCO terminated prepolymer, synthesis was controlled by determing the NCO groups by the dibutyl amine back-titration method until a theoretical value was achieved. The theoretical value of NCO groups for PUD based on oligoester polyol obtained from glycolysis with molar ratio of PET repeating unit to glycol, 1:10, was 14.2% for PG, 13.7%, for TEG and 13.9% for poly(ethylene glycol). The neutralization of the carboxylic groups in hydrophilic monomer, extension chain with EDA and the preparation of stable dispersion was done as in the previous modified acetone process. The solid content in this dispersion was 30%.

Films were prepared by casting the aqueous dispersions on the glass surface and allowing them to dry at room temperature for 7 days and then at 60 °C, for 12h (Coutinho, 1996, 2003). Films were cast by 100 μm applicators from the solutions onto glass surface (7cm x 2 cm) to obtain dry film thickness of 25-30 μm, making demoulding impossible. After demoulding, the films were kept into a desicator to avoid moisture. TG experiments were performed in a Perkin-Elmer TG-7 analyser. Film samples about 20 mg were placed in a platinum sample pan and heated from 30 to 600 °C, with an air flow of 200 mL min[-1] and heating rates of 10 °C min[-1]. During the heating period, the weight loss and temperature difference were recorded as a function of temperature.

TGA was used to analyze decomposition behavior of cured films of PUD synthesized with glycolized products obtained from PET waste. TG curves are depicted in Figs. 4a,4c and 4e (curve marked as 1 a show lower molar ratio of PET / glycol (1:2) in the glycolized oligoester) represents the degradation of PUD influenced by the variation of oligoester polyols. It was verified that the thermal stability was influenced by glycol type and different molar ratio of PET repeating unit to glycol in glycolysis reaction. Figs. 4b, 4d and 4f depict the behavior of corresponding differential weight loss (DTG) curves.

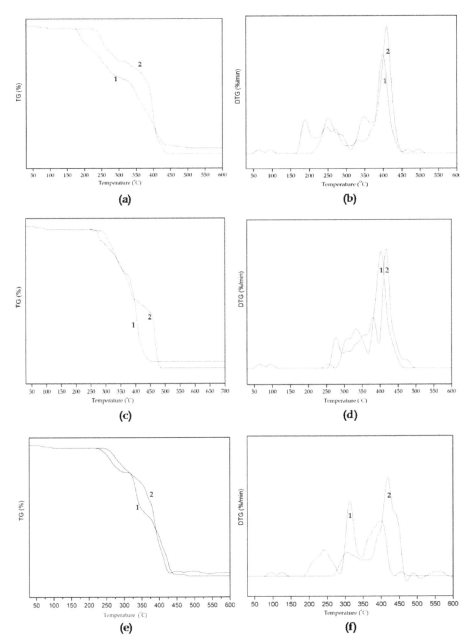

Figure 4. TG curves (a) and DTG curves (b) of PUD synthesized from glycolized oligoester PET/PG with molar ratio 1:2 (1) and 1:10 (2). TG curves (c) and DTG curves (d) of PUD synthesized from glycolized oligoester PET/TEG with molar ratio 1:2 (1) and 1:10 (2). TG curves (e) and DTG curves (f) of PUD synthesized from glycolized oligoester PET/poly(ethylene) glycol 400 with molar ratio 1:2 (1) and 1:10 (2).

PUD (synthesized from glycolized oligoester PET/PG (1:2), Fig. 4b) synthesized from depolymerised oligoesters with lower molar ratio of PET repeating unit to glycol in glycolysis reaction showed lower thermal stability in the initial stage of degradation may be due to the presence of greater amount of aromaticity in polyester backbone which makes the PU chains susceptible to scission and relieves the structure crowing (Athawale & Kulkarni, 2010). In later stage (above 300 °C), it showed enhanced thermal stability. It has also been proved that two or three peaks of first decomposition were well correlated with the higher value of polydispersity of GPC results (1.65), for oligoester polyols PET/PG (1:2) compared to the values of polydispersity (1.20), for oligoester polyols PET/PG(1:10). Curve marked as 2, in Figures 4b, 4d and 4f, which shifted third decomposition step temperature to the higher values, shows that glycolized oligoester obtained with higher molar ratio of PET/glycol of 1:10 have better thermal stability of obtained PUD.

Because of the presence of oligoester polyols wich lower molecular weight in glycolysis reaction and a diamine were used in the synthesis of PUD, two kinds of hard segments are formed, urethane and urea. It has been estabilished that the urethanes have lower thermal resistance than urea and therefore the first decomposition process at about 190 - 250 °C and the second at about 270 – 290 °C of PUD should correspond to the urethane and urea hard segments, respectively. The decomposition temperature of the soft segment is observed at 400-430 °C. The decomposition temperature for investigated samples are listed in Table 1.

Sample	First decomposition		Second decomposition	Third decomposition
	T_{1on} (°C)	T_{1max} (°C)	T_2 (°C)	T_3 (°C)
PET/PG (1:2)	190.2 247.0	270.7	349.7	400.3
PET/TEG (1:2)	-	288.0	335.1	401.9
PET/PEG$_{400}$ (1:2)	248.6	-	316.5	403.4
PET/PG (1:10)	251.7	288.1	327.0	411.3
PET/TEG (1:10)	-	277.6	379.6	416.4
PET/PEG$_{400}$ (1:10)	-	-	305.4	422.4

Table 1. Temperature of decomposition of PUD

The degradation profile of PUD was dependent on mole ratios of PET to glycol in glycolyzed products.

The samples based on PET/glycol, at molar ratio of 1:10, had better thermal stability than samples based on PET/glycol, at molar ratio of 1:2. The higher values of temperature for third decomposition stage, for samples with molar ratio of 1:10, probably is due to the increased length of glycol in glycolyzed oligoester polyol.

6. Waterborne PUD based on polycarbonate diols (PCD)

The polyols used in PUD synthesis are of polyether-, polyester-, polycaprolactone- and polycarbonate- origin. The use of individual types of polyol chains and their functionalities depend on the purpose of the potential use, e.g.; PUD made from polyesters can have slightly elevated strength and oil resistance compared to polyether-based PUD and have been largely used in PU paints as they exhibit outstanding resistance to light and aging. Polyether polyols are susceptible to light and oxygen when hot, however, they improve water dispersion, and impart chain flexibility (Gunduz & Kisakurek, 2004). The use of PCD in PUD, as compared to other polyols, imparts better hydrolysis resistance, improved ageing and oil resistance, excellent elastomeric properties even at low temperature, improved mechanical properties, good weathering and fungi resistance (Garcia et al., 2010). PCD used as the soft segment component in PUD synthesis are usually obtained from dimethylcarbonate or ethylene carbonate and a linear aliphatic diol (Foy et al., 2009). The properties of PUD are related to their chemical structure (Cakić et al., 2009; Athawale & Kulkarni, 2010) and are mainly determined by the interactions between the hard and soft segments, and the interactions between the ionic groups (Garcia et al., 2011). The properties of PUD are strongly influenced by composition and ionic content, an important target in an investigation of the role of the composition (Lee at al., 2004, 2006).

6.1. Experimental

Water-based PUD derived from IPDI, with different molar ratio PCD to DMPA, were prepared by the modified dispersing process. The ionic group content in PU-ionomer structure was varied by changing the amount of the internal emulsifier, DMPA (4.5, 7.5 and 10 wt% to the prepolymer weight).

Three waterborne PUD were prepared using NCO/OH = 1.5 by method in which the dispersing procedure was modified (Lee et al., 2006). In the modified procedure only the dispersing stage was varied compared to the standard procedure. The prepolymer solution was mixed with a small amount of deionized water for dispersion of polymer in water. Solvent was added for reducing the viscosity, if necessary.

Into a 250 ml glass reaction kettle, equipped with a mechanical stirrer containing a torque meter, a thermometer, a condenser for reflux and nitrogen gas inlet, was added 60 g (0.03 mol) of PCD (dried under vacuum at 120 °C); and 4, 8 or 12 g (0.03, 0.06 or 0.09 mol) of DMPA dispersed in 30 ml DMF. The reaction mixture was heated at 70 °C for 0.5 h to obtain a homogeneous mixture. This step is important for the resulting equal uniform distribution of hydrophilic monomer, DMPA, on PU backbone. After that 20, 32 or 40 g (0.09, 0.15 or 0.18 mol) of IPDI and DBTDL (0.03 wt. % of the total solid) were added to the homogenized mixture and stirred at 80 °C for 2.5 h. Dibutyl amine back titration method was used for the determination of the reaction time necessary to obtain completely NCO-terminated prepolymer. Then the mixture was cooled down to 60 °C and carboxylic groups (DMPA equiv) were neutralized with TEA dissolved in NMP (2 wt % of the total reaction mass) by stirring the reaction mixture at 60 ˚C for 1h.

Subsequently, the prepolymer solution was mixed with 0.3 ml of deionised water for dispersion step-by-step. Stirring was increased during the addition of water, and the mixture was diluted with NMP. Waterborne PUD was obtained by drop-wise addition of water to the mixture in order to obtain PUD with solid content of about 30% at 30 °C for 1h. The chain extension was carried out with solution of 0.9 or 1.8 g (0.015 or 0.03 mol) of EDA in 2 ml of deionised water at 35 °C for 1h. The mixture was heated to 80 °C under vacuum in order to remove NMP and to obtain PUD with solid content of about 30%.

The thermal stability of PU was determined by TG. The DTG thermogram of cured films of PUD based on PCD showed several degradation steps (Fig.5b). Detailed analysis of the thermogram is represented in Table 2. The decrease in the DMPA content produces a slight increase in the decomposition temperature (Fig.5a). However, the degradation mechanism was very complex due to the different stability of the hard and soft segments.

The removal of residual water due to incomplete drying of PU was produced at around 130 °C. The DTA thermogram of the used aliphatic PCD shows the main degradation at 350 °C and other less important at 265 °C (Garcia, 2010, 2011). Because this diol and EDA were used in the synthesis of PU, two kinds of groups in hard segments have been formed, i.e., urethane and urea ones. The decomposition temperature of PUD is mostly influenced by the chemical structure of the component having the lowest bond energy (Coutinho et al., 2003; Cakic et al.,2009). The urethane bond has lower thermal resistance than the urea bond and thus the first decomposition process at about 280 °C corresponds to the beginning of the urethane part of hard segment degradation and second at about 300 °C to the degradation of the urea part of hard segment. The degradation in PU at 264–268 °C is characteristic of the polyol. The degradation of soft segment (mainly composed of polyol) is produced at 329–338 °C. The soft and hard segments content were quantified from the weight loss at above mentioned temperatures (265 °C from PCD degradation, 280 °C and 300 °C correspond to the urethane and urea hard segment degradation, and 330 °C from degradation of soft segment).

According to Table 2, the decrease in DMPA content produced a slight increase in the decomposition temperature and a decrease in the weight loss for the decomposition of urethane and urea hard segments, which can be ascribed to a decrease in the amount of hard segment. The decomposition temperature of the soft segments is produced at 329–338 °C and the weight loss increase by decreasing DMPA content 20.9 and 23.5 to 6.2wt.% in PU ionomers.

Sample PUD	Residual water		Soft segment		Hard segment	
	T (°C)	Weight loss (wt%)	T (°C)	Weight loss (wt%)	T (°C)	Weight loss (wt%)
4.5%DMPA	129.8	1.1	268; 329	10.1; 20.9	282; 301	23.5; 45.4
7.5%DMPA	137.3	1.0	264; 338	3.8; 23.5	290; 304	17.5; 53.7
10%DMPA	128.8	0.9	267; 331	10.5; 6.2	280; 301	22.9; 59.3

Table 2. Temperature of decomposition and weight loss of PUD (obtained by TG measurements)

The decrease in DMPA content produces a decrease in the hard segment content of PU ionomers. The resistance to thermal degradation of PU ionomer increased by decreasing the DMPA content due to the lower hard segment content.

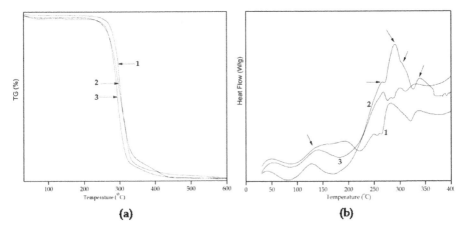

Figure 5. TGA curves (a) and DTA curves (b) of cured films of PUD based on PCD with 4.5% DMPA (1), 7.5% DMPA (2), 10% DMPA (3).

7. Conclusions

The wide application of PUD makes necessary better understanding of the chemistry-structure relationship that improves the thermal stability as this is important prerequisite to obtain tailor-made products for high performance applications. In this work, the investigation on thermal degradation of PUD with well-defined architectures indicated that diol types and DMPA content had great influence on thermal stability. PUD with lower DMPA content has shown enhanced thermal stability. The degradation profiles of PU aqueous dispersions were influenced by the type of chain extender, length of the hard segment and type of catalysts. The TG curves showed that the length of the hard segment had a strong influence on the thermal profile of the samples as a whole. The possibility for using glycolysis products of waste PET in PUD manufacturing was confirmed. The effects of glycol type and the different mole ratios of PET to glycol on thermal properties of PUD have been described. The degradation profile of the dispersions was dependent on mole ratios of PET to glycol in glycolyzed products. The samples based on (PET/glycol molar ratio 1:10) have shown enhanced thermal properties, which can be ascribed to increased length of glycol in glycolyzed oligoester polyol.

Author details

Suzana M. Cakić and Olivera Z. Ristić
University of Niš, Faculty of Technology, Leskovac, Serbia

Ivan S. Ristić
University of Novi Sad, Faculty of Technology, Novi Sad, Serbia

8. References

[1]   Atta A.M.; Abdel-Raouf M.E.; Elsaeed S.M.; Abdel-Azim A.A. (2006) Curable resins based on recycled poly(ethylene terephthalate) for coating applications. *Prog. Org. Coat.,* 55, pp. 50–59.

[2]   Atta A.M.; El-Kafrawy A.F.; Aly M.H.; Abdel-Azim A.A. (2007) New epoxy resins based on recycled poly(ethylene terephthalate) as organic coatings. *Prog. Org. Coat.,* 58, pp. 13–22.

[3]   Athawale V.D. & Kulkarni M.A. (2009) Preparation and properties of urethane/acrylate composite by emulsion polymerization technique. *Prog. Org. Coat.,* 65, pp. 392–400.

[4]   Athawale V.D. & Kulkarni M.A. (2010) Polyester polyols for waterborne polyurethanes and hybrid dispersions. *Prog. Org. Coat.* 67, pp. 44–54.

[5]   Bechara I. (1998) Formulating with polyurethane dispersions. *Eur. Coat. J.,* 4, pp. 236–243.

[6]   Blank W.J.; He Z.A. & Hessell E.T. (1999) Catalysis of the isocyanate-hydroxyl reaction by non-tin catalysts. *Prog. Org. Coat.,* 35, pp. 19-29.

[7]   Blank W.J. & Tramontano V.J. (1996) Properties of crosslinked polyurethane dispersions. *Prog. Org. Coat.,* 27(1), pp. 1-15.

[8]   Cakić S. M.; Lačnjevac Č.; Stamenković J.; Ristić N.; Takić Lj.; Barać M.; Gligorić M. (2007 a) Effects of the acrylic polyol structure and the selectivity of the employed catalyst on the performance of two-component aqueous polyurethane coatings. *Sensors,* 7(3), pp. 308-318.

[9]   Cakić S.M.; Nikolić G.S.; Stamenković J.V. (2007 b) Thermo-oxidative stability of waterborne polyurethanes with catalysts of different selectivity evaluated by non-isothermal thermogravimetry. *J. Serb. Chem. Soc.,* 72(7), pp. 723-735.

[10] Cakić S.M.; Nikolić G.S.; Lačnjevac Č.; Gligorić M.; Rajković M. (2007 c) The thermal degradation of aqueous polyurethane with catalysts of different selectivity. *Prog. Org. Coat., 60(2),* pp. 112-116.

[11] Cakic S.; Lačnjevac Č.; Rajković M.B.; Rašković Lj.; Stamenković J. (2006 a) Reticulation of Aqueous Polyurethane Systems Controlled by DSC Method. *Sensors,* Vol. 6, No. 5, pp. 536-545.

[12] Cakic S.; Lačnjevac Č.; Nikolić G.; Stamenković J.; Rajković M.B.; Gligorić M.; Barać M. (2006 b) Spectroscopic Characteristics of Highly Selective Manganese Catalysts in Acqueous Polyurethane Systems. *Sensors,* 6, pp. 1708-1720.

[13] Cakić S.; Ristić I.; Djordjević D.; Stamenković J.; Stojiljković D. (2010) Effect of the chain extender and selective catalyst on thermooxidative stability of aqueous polyurethane dispersions. *Prog. Org. Coat.,* 67, pp. 274–280.

[14] Cakić S.; Ristić I.; M-Cincović M.; Nikolić N.; Ilić O.; Stojiljković D.; B-Simendić J. (2011) Glycolyzed products from PET waste and their application in synthesis of polyurethane dispersions. *Prog. Org. Coat.,* In press, doi: 10.1016/j. porgcoat.2011.11.024.

[15] Cakić S.; Stamenković J.; Djordjević D.; Ristić I. (2009) Synthesis and degradation profile of cast films of PPG-DMPA-IPDI aqueous polyurethane dispersions based on selective catalysts. *Polym. Degrad. Stab.,* 94, pp. 2015–2022.

[16] Chang T.C.; Chiu Y.S.; Chen H.B.; Ho S.Y. (1995) Degradation of phosphorus-containing polyurethanes. *Polym. Degrad. Stab.*, 47, pp. 375-381.

[17] Chattopadhyay D.K. & Webster D.C. (2009) Thermal stability and flame retardancy of polyurethanes. *Prog. Polym. Sci.* 34, pp. 1068-1133.

[18] Collong W.; Gobel A.; Kleuser B.; Lenhard W.; Sonntag M. (2002) 2K waterborne clearcoat-a competition between crosslinking and side reactions. *Prog. Org. Coat.*, 45, pp. 205-209.

[19] Coutinho F.M.B. & Delpech M.C. (1996) Some properties of films cast from polyurethane aqueous dispersions of polyether-based anionomer extended with hydrazine. *Polym. Test.*,15, pp. 103-113.

[20] Coutinho F.M.B.; Delpech M.C.; Alves T.L.; Ferreira A.A. (2003) Degradation profiles of cast films of polyurethane and poly(urethane-urea) aqueous dispersions based on hydroxyterminated polybutadiene and different diisocyanates. *Polym. Degrad. Stab.*, 81, pp. 19-27.

[21] Delpech M.C. & Coutinho F.M.B. (2000) Waterborne anionic polyurethanes and poly(urethane-urea)s: influence of the chain extender on mechanical and adhesive properties. *Polym. Test.*, 19, pp. 939-952.

[22] Dieterich D. (1981) Aqueous Emulsions, Dispersions and Solutions of Polyurethanes; Synthesis and. Properties. *Prog. Org. Coat.*, 9, pp. 281-340.

[23] Dulog L. & Storck G. (1966) Die oxydation von polyepoxiden mit molekularem sauerstoff, *Macomol. Chem.*, 91, pp. 50-73.

[24] Fambri L.; Pegoretti A.; Gavazza C.; Penati A. (2000) Thermooxidative Stability of Different Polyurethanes Evaluated by Isothermal and Dynamic Methods. *J. Appl. Polym. Sci.*, 81, pp. 1216–1225.

[25] Foy E.; Farrell J.B.; Higginbotham C.L. (2009) Synthesis of Liner Aliphatic Polycarbonate Macroglycols Using Dimethylcarbonate. *J. Appl. Polym. Sci.*, 111, pp. 217–227.

[26] Garcia-Pacios V.; Costa V.; Colera M.; J. Martin-Martinez M. (2010) Affect of polydispersity on the properties of waterborne polyurethane dispersions based on polycarbonate polyol. *Int. J. Adhes. Adhes.*, 30, pp. 456–465.

[27] Garcia-Pacios V.; Costa V.; Colera M.; Martin-Martinez J.M. (2011) Waterborne polyurethane dispersions obtained with polycarbonate of hexanediol intended for use as coatings. *Prog. Org. Coat.*, 71, pp. 136–146.

[28] George Woods, (1987). *The ICI Polyurethanes Book*, 2nd Edition, Wiley, New York.

[29] Gunduz G. & Kisakurek R.R. (2004) Structure–Property study of waterborne polyurethane dispersions with Different hydrophilic content and polyols. *J. Disper. Sci.Technol.*, 25 (2), pp. 217-228.

[30] Hepburn C. (1992) *Polyurethane Elastomers*, Second ed., Elsevier, New York.

[31] Jacobs, P.B. & Yu, P.C. (1993) Two-Component Waterborne Polyurethane Coatings. *J.Coat. Tech.*, 65 (822), pp. 45-50.

[32] Jang J. Y.; Jhon Y.K.; Cheong I.W.; Kim J.H. (2002) Effect of process variables on molecular weight and mechanical properties of water-based polyurethane dispersion. *Colloids Surf. A- Physicochem. Eng. Aspects*, 196, pp. 135-143.

[33] Kim B.K. & Min L.Y. (1994) Aqueous dispersion of polyurethanes containing ionic and nonionic hydrophilic segments. *J. Appl. Polym. Sci.*, 54, pp. 1809-1815.

[34] Kim B. K. (1996) Aqueous polyurethane dispersions. *Coll. Polym. Sci.*, 274, pp. 559-611

[35] Lee D.K.; Tsai H.B.; Tsai R.S. (2006) Effect of Composition on Agueous polyurethane Dispersions Derived from polycarbonatediols. *J. Appl. Polym. Sci.*, 102, pp. 4419–4424.

[36] Lee D.K.; Tsai H.B.; Wang H.H.; Tsai R.S. (2004) Aqueous Polyurethane Dispersions Derived from Polycarbonatediols. *J. Appl. Polym. Sci.*, 94, pp. 1723–1729.

[37] Lee H.T.; Hwang Y.T.; Chang N.S.; Huang C.C.T.; Li H.C. (1995) Waterborne, High-Solids and Powder Coatings Symposium, New Orleans, pp. 224.

[38] Lu M.G.; Lee J.Y.; Shim M.J.; Kim S.W. (2002) Thermal Degradation of Film Cast from Aqueous Polyurethane Dispersions. *J. Appl. Polym. Sci.*, 85, pp. 2552–2558.

[39] Oertel G. (1985) *Polyurethane Handbook*, Hanser Publishers, Munich.

[40] Patel M.R.; Patel J.V.; Mishra D.; Sinha V.K. (2007) Synthesis and Characterization of Low Volatile Content Polyurethane Dispersion from Depolymerised Polyethylene Terphthalate. *J. Polym. Eviron.*, 15, pp. 97–105.

[41] Pielichowski K.; Slotwinska D.; Dziwinski E. (2004) Segmented MDI/HMDI based polyurethanes with lowered flammability. *J. Appl. Polym. Sci.*, 91, pp. 3214-3224.

[42] Prime R.B.; Burns J.M.; Karmin M.L.; Moy C.H.; Tu H.B. (1988) Applications of DMA and TGA to quality and process control in the manufacture of magnetic coatings. *J. Coat. Technol.*, 60, pp. 55–60.

[43] Ramesh S. & Radhakrishna G. (1994) Synthesis and characterization of polyurethane ionomers. *Polym. Sci.*, 1, pp. 418-423

[44] Randell D. & Lee S. (2000) *Polyurethane Book*, 2nd Editors, John Wiley & Sons, New York.

[45] Rothause J. W. & Nechtkam K. (1987) Advances in Urethane Science and Technology, 10, pp. 121

[46] Rosthauser J.W. & Nachtkamp K.J. (1986) Waterborne polyurethanes. *J.Coat. Fabrics.*, 16, pp. 39-79.

[47] Scaiano J. C. (1989) Laser Photolysis in Polymer Chemistry. Degradation and Stabilization of Polymers. Elsevier, Amsterdam.

[48] Stamenković J.; Cakić S.; Nikolić G. (2005) Study of the catalytic selectivity of an aqueous two-component polyurethane system by FTIR spectroscopy. *Chem. Ind.*, 57, pp. 559-562.

[49] Tawa T. & Ito S. (2006) The Role of Hard Segments of Aqueous Polyurethane-urea Dispersion in Determining the Colloidal Characteristics and Physical Properties. *Polym. J.*, 38(7), pp. 686-693.

[50] Wang T.L. & Hsieh T.H. (1997) Effect of polyol structure and molecular weight on the thermal stability of segmented poly(urethaneureas). *Polym. Degrad. Stab.*, 55, pp. 95-102.

Bio-Based Polyurethanes

Polyglucanurethanes: Cross-Linked Polyurethanes Based on Microbial Exopolysaccharide Xanthan

Nataly Kozak and Anastasyia Hubina

Additional information is available at the end of the chapter

1. Introduction

Considering environmental protection and resolution a number of ecological problems (including problem of recourses for chemical synthesis depletion) synthesis of the biodegradable polymer materials becomes one of the most actual tasks of modern polymer chemistry. Among ways of environmental protection from polymer waste (keeping on waste deposits, burials, incineration, pyrolysis, recycling) there can be distinguished the method of minimization of ecological pollution due to creation of polymers able to be destructed under influence of natural factors such – chemical (oxygen, air, water), physical (sun light, heat), biological (bacteria, fungi) etc. These factors are very effective and lead to fragmentation of polymer as a result of macromolecule degradation and turning it into low-molecular compounds which become part of natural circuit of substance. In other words biodestruction is reliable and comparatively fast method of utilization. Usually it can be achieved by implication of natural compounds fragments into polymer structure. Other promising method is biopolymers modification with further creation of new synthetic polymers able for degradation under biological factors. Development of this method in future allows to resolve one of the most actual modern problems and to substitute petroleum refining products as the base of chemical synthesis with renewable source. It is also relevant using as reagents economically effective products which are cheaper than oil refining raw materials.

Purpose of our study was to create new polymerizing systems possessing above metioned attractive features. Therefore new polyglucanurethane (PGU) networks were obtained on the base of microbial polysaccharide xanthan and blocked polyisocyanate (PIC) using environment friendly method. Biopolymer application as reagent provides both preserving advantages of initial materials and developing new advanced properties of obtained biodegradable materials due to chemical modification. Replacement of toxic compounds with latent isocyanate-containing reagent blocked polyisocyanate is also a prominent advantage of developed technique. PGU were obtained via interaction of xanthan hydroxyl groups and isocyanate groups of deblocked above 125°C PIC (Kozak & Nizelskii, 2002).

Microbial polysaccharide xanthan is produced by xanthomonas campestris pv. Campestris bacteria (Gzozdyak et al., 1989). Xantan is well known and most widely used microbial exopolysaccharide. It is used in light industry (textile), heavy industry (drilling and oil production) and food industry as well as in agriculture, forestry, pharmaceutics, medicine and cosmetics. The water solutions of xantane have unique rheological properties due to metal cation complexing ability and formation of primary, secondary and higher levels of structure. The main chain of the polysaccharide is alike to cellulose and its side-chains (pendants) consist of glucose, mannose and glucuronic acid residues.

Blocked polyisocyanate is latent reagent which is able to produce reactive isocyanate groups under elevated temperature. It is multifunctional latent reagent that can consist of 40 to 70 % of dimeric compound , 20 to 60 % of tetramer and 1 to 5 % of trimer and hexamer. Melting temperature interval of blocked PIC is from 80 to-95 °C, NCO-group unblocking temperature ranges from 125 to 130 °C. PIC is soluble in most of organic solvents and can be used both in powder and liquid form.

2. Synthesis

The reaction was provided in solid. Fig. 1 describes the scheme of PGU synthesis. There occur both deblocking of polyisocyanate groups and urethane bonds formation.

There were obtained powdered PGU, hot-pressed samples and reinforced PGU with calculated degree of polysaccharide hydroxyl group substitution of 20, 40, 60, 80 and 100%. As far as obtained polymer is quite new and unexplored polymer material, not full range of the hydroxyl/urethane ratio was studied by methods presented in this Chapter. The obtained materials are acid-, alkali- and thermo resistant.

The reaction path was controlled using sampling procedure and sample analysis with FTIR spectroscopy.

Characteristic band at 2276 cm^{-1} which appears after heating the reactive mixture up to 130°C demonstrates the process of isocyanate groups deblocking. On the initial stages of reaction all of PGU samples show increasing intensity of this band due to active isocyanate group formation.

Interaction of N=C=O groups with the nearest primary and secondary hydroxyl groups of polysaccharide leads to decreasing intensity of characteristic band 2276 cm^{-1} during next 10 min. Diffusion limitations of this reaction are determined by heterogeneity of reactive mixture and results in retarding of urethane bonds formation. Mechanic stirring of reactive mixture allows improve reactive centers contact and leads to total disappearance of isocyanate groups in the system. (Fig 2.a). Consumption of hydroxyl groups is accompanied by disappearance of the band at 3165 cm^{-1} and decreasing of intensity band at 1209 cm^{-1} (valence vibrations and deformation vibrations of O-H bond in glucuronic acid residue respectively) and by intensity reduction of the band at 3215 cm^{-1} (valence vibrations of primary OH-groups in mannopyranose cycles) (Fig 2 c).

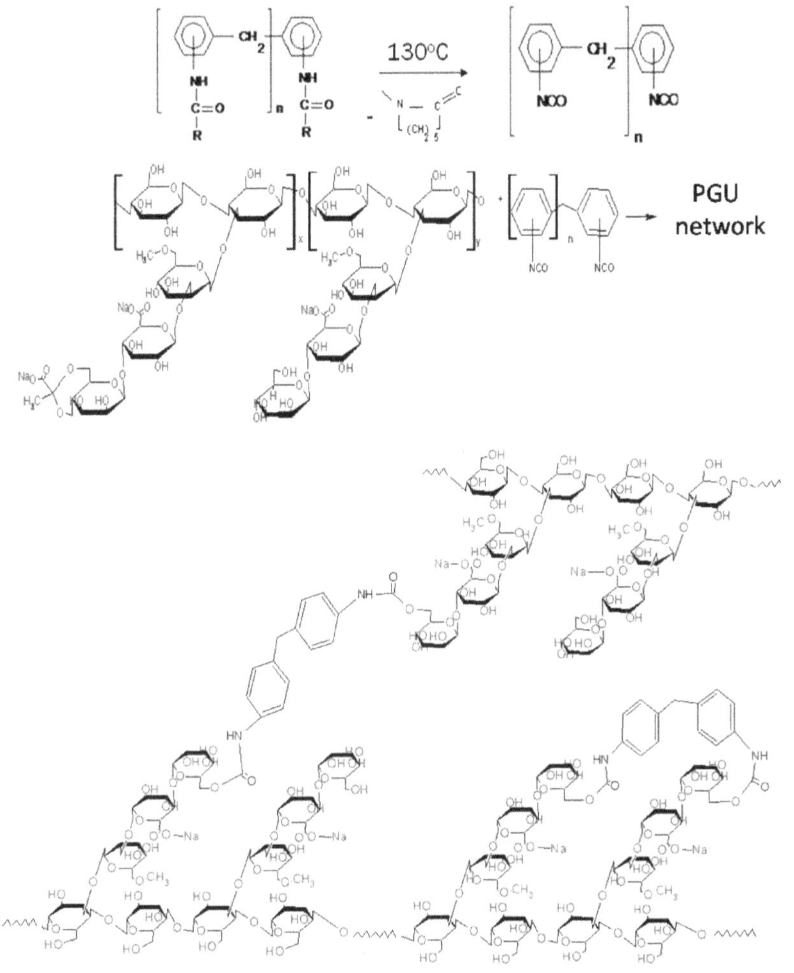

Figure 1. Scheme of PGU synthesis

Increasing intensity of the band at 3364 cm^{-1} (in characteristic doublet of N-H valence vibrations), appearance of the 1635 cm^{-1} band in the region of NH deformation vibrations (amide II) and changes of intensity of 1650 and 1590 cm^{-1} bands respond to formation of urethane bonds and releasing of blocking agent (Fig 2b).

In the wave numbers range from 3000 to 3500 cm^{-1}redistribution is observed of the intensities of absorption bands corresponding to hydrogen linked OH-groups. That points on redistribution of intermolecular bonds in the system during polysaccharide cross-linkage and PGU formation.

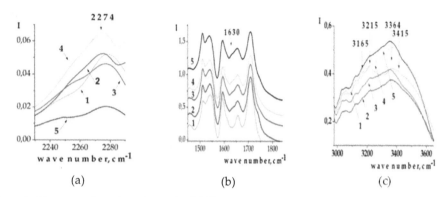

Figure 2. IR spectra of reactive mixture for PGU60 at temperature 130°C(1), 150°C (2), at 150°C after 10 min (3), after 20 min (4) after 30 min(5)

According to FTIR spectra of various PGU (PGU40, PGU80 and PGU100) the number of isocyanate groups released at the initial stage of reaction correlates with the polyisocyanate content in the system. During the first 10 min the process of polyisocyanate deblocking dominates. At the same time the urethane group formation occurs via interactiom of NCO groups and polysaccharide hydroxyl groups in acid residue of glucuronic acid and/or hydroxyl groups of mannose. The time when the urethane group formation begins to dominate depends on balance of the reagents in reaction mixture.

3. Polyglucanurethane chemical and thermal resistance

Obtained PGU networks possess advanced thermal and chemical (both alkali and acid) resistance. Chemical resistance of PGU was analyzed using standard method...[State Standart 12020-72]. Thermal resistance of initial reagents and PGU of various composition were studied with the thermogravimetry. Table 1 shows the results of PGU20 exposure in water, acid- and alkali media. Fig 4. illustrates the TGA curves: mass loss (TG), differential mass loss (DTG) and differential thermal analysis (DTA) that characterize the dependence of thermooxidative destruction character of PGU on the degree of substitution of xanthan hydroxyl groups.

As it can be seen from table 1 data the mass of PGU20 samples remains practically unchanged after the 7 and 13 days exposure in aggressive alkali end acid medium. The initial stages of PGU interaction with water, alkali- and acid media are characterized with ignificant swelling of polysaccharide component.

According to TGA data thermooxidative destruction of the systems analyzed consists of several stages. TGA curves of xanthan are typical for polysaccharides. TGA curves character for PGU networks and number of stages in temperature interval from 20 to 700°C depend on the balance of hydroxyl and urethane groups in PGU. Stage of weight loss in temperature interval 60-140°C is accompanied with endothermal peak on DTA curve and responds to absorbed water removal.

Time,	Sample weight, g		
day	H_2O	H_2SO_4	$NaOH$
1	0,081	0,085	0,081
2	0,213	0,205	0,431
3	0,217	0,213	0,489
7	0,250	0,224	0,485
13	0,425	0,241	0,489

Table 1. The weight change of PGU20 exposed in deionized water, concentrated sulphuric acid (V=20 ml, 30% wt.) and concentrate alkali solution (V=20 ml, 40% NaOH).

Presenting mass of absorbed water in modified and non-modified samples of exopolysaccharide as mass loss at the first stage (temperature interval 45 – 150°C) we can see that amount of absorbed water correlates with balance of hydroxyl and urethane groups in the system (Fig.3). It corresponds with the fact that system hydrophilic properties correlate with amount of hydroxyl groups. Weight loss at this stage is 1,5; 2,5; 8,5 % wt. for PGU80, PGU40, PGU20, respectively.

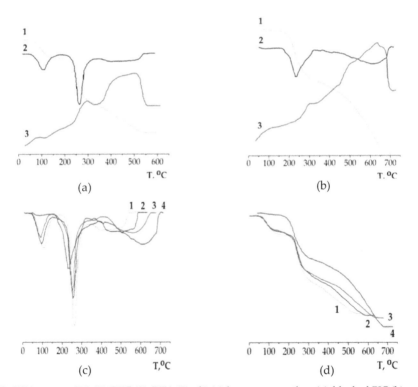

Figure 3. TGA curves - TG (1), DTG (2), DTA (3) of initial reagents: xanthan (a), blocked PIC (b); curves DTG (c) and TG (d) for initial reagents and PGU of various composition: xanthan (1), 2- PGU20 (2), PGU40, PIC (4)

Intensive thermal-oxidative destruction is observed in 200 – 400°C temperature interval. The characteristic temperature responding to maximum speed of weight loss at the stage shifts toward the higher temperatures with increasing of urethane groups amount in the system. Weight loss at this stage is 35, 30, 35 % wt. for PGU80, PGU40, PGU20, respectively.

Wide peak on DTG curves in temperature interval from 500 to 700°C mainly corresponds to destruction of carbon base of polymer. With growth of urethane group content in the system the carbon residual decreases.

As it can be seen thermal resistance of PGU grows with increasing of urethane group amount. For instance, for PGU40 high temperature stage is allocated 40°C lower than for the PGU 80. The degree of hydroxyl substitution also influences the system capacity of water absorbance. With growth of urethane group content in the system the amount of absorbed water declines

4. PGU interaction with water solutions of phenol and transition metal salts

It is known that microbial polysaccharides are considered as prospective raw materials for obtaining effective sorbents for extracting organic compounds and metal ions from water solution (Crini, 2005). Application of water-soluble polysaccharides (eg. xanthan) as sorbents is difficult. Using of PGU allows both keep sorption properties of polysaccharide and eliminate a number of disadvantages (water solubility, low chemical resistance etc.). Study of properties of cross-linked PGU (Hubina, 2009) revealed its ability to quantitative extraction of phenol and bivalent metal ions from water solution, while controlling capacity of the material with cross-linking degree.

To analyze ability of PGU to phenol sorption from its water solution the pollutant concentration was controlled before and after exposure of PGU20 and PGU60 films in 10^{-4}M phenol water solution during 24, 48 and 120 hours. Phenol concentration was monitored using UV electron spectroscopy by the change of band intensity near 256 nm. Experiment conditions were as follows: T=18 °C, $m_{sorbent}$=7 g, $V_{solution}$=50 ml. Calibration curve was plotted for 0,05; 0,075; 0,1; 0,15 and 0,2 M phenol water solutions.

According to UV-spectroscopy data after 24 hours of exposure PGU in phenol solution increasing of phenol concentration is observed both for PGU20 and PGU60 (Fig.4). Further exposure of PGU in solution leads to essential decreasing of phenol concentration. Such phenol concentrating during PGU interaction with phenol solution can be explained by the preferable swelling of polysaccharide component of the polymer in water.

The same character of concentration change is observed while extracting Cu^{2+} ions from copper sulfate solution. (Fig.5). Its concentration change was controlled using electron spectroscopy in Vis-region by the change in band intensity near 810 nm. That band corresponds to d-d transitions in $[Cu(H_2O)6]^{2+}$aqua ion. Exposure of PGU60 sample in copper sulfate water solution during 10 min. is accompanied by sufficient growth of solution absorbance and shows copper ions concentrating. Increasing time of interaction with sorbent to 60 min leads to absorbance reduction.

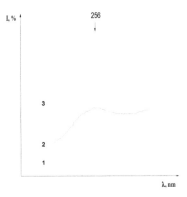

Figure 4. The electron spectra of phenol water solutions: 1 – initial; 2 – PGU60 after 10 min exposure; 3 – PGU20 after 10 min exposure

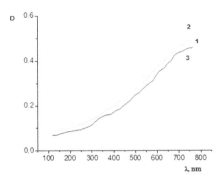

Figure 5. The electron spectra of copper salt water solutions: 1 - initial, 2 - after 10 min of PGU exposure and 3 - after 60 min of PGU exposure

Interaction of PGU40 and PGU80 with water solution of Cu^{2+} and Co^{2+} salts with concentration of 50 and 500 mg/dm³ was studied in static conditions using conductometry. Solution conductivity was fixed after 2, 12 and 74 hours of PGU exposure. Experiment conditions were as follows m_{PGU}=1,25 g; $V_{solution}$=50 cm³ ; permanent stirring frequency=2 Hz; room temperature. Intermediate and final concentrations were calculated from calibration curve. Fig 7 demonstrates conductivities of cobalt and copper salt solutions of various concentrations (50 and 500 mg/dm³) depending on the time of interaction with PGU40 and PGU80. In the table 2 there are the results of concentration changes of copper(2+) sulfate and cobalt(2+) chloride solutions respectively *vs* time of exposure PGU40 and PGU80. Fig 6 shows that conductivity of cobalt chloride and copper sulfate solutions grows on the initial stages of sorbent exposure for both high (500 mg/dm) and low (50 mg/dm) concentrations.

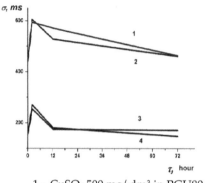
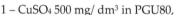
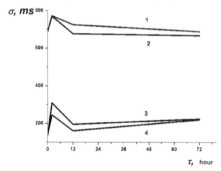

1 – CuSO₄ 500 mg/ dm³ in PGU80,
2 - CuSO₄ 500 mg/ dm³ in PGU40,
3 - CuSO4 50 mg/ dm³ in PGU80,
4. - CuSO₄ 50 mg/ dm³ in PGU40

1 - CoCl₂ 500 mg/ dm³ in PGU80,
2 - CoCl₂ 500 mg/ dm³ in PGU40
3. - CoCl₂ 50 mg/ dm³ in PGU80,
4. - CoCl₂ 50 mg/ dm³ in PGU40

Figure 6. Dependence of solution conductivity *vs* time of sorbent exposure

Similar effect was observed for hot-pressed samples of PGU while interacting with water solutions of phenol and transition metal salts due to predominant swelling of polysaccharide component on the initial stages. Next 12 and 74 hours of PGU exposure result in sufficient solution concentration decreasing due to metal ions capture by the functional groups of sorbent. Equilibrium was achieved both for PGU40 and PGU80 in solutions of high cobalt salt concentration (500 mg/ dm³) while for copper salt solutions equilibrium was achieved for low concentration (50 mg/ dm³).

As table 2 and Fig 6 show the efficiency of metal ions extraction depends on both concentration of solution and the balance of hydroxyl and urethane groups in PGU. The highest concentration fall was observed for PGU80. The solutions of low concentration aren't sensible to hydroxyl and urethane balance in PGU. Besides, the best ion extraction of cobalt ions from low concentration solutions is achieved with short-time exhibition of sorbent. For cobalt salts solutions of high concentration and copper salts solutions of low concentration the best effect is achieved during 74 hours of exhibition.

PGU	C, mg/dm³	Cu SO₄			CoCl₂		
		ΔC_{2h} , %	ΔC_{12h} , %	ΔC_{74h} , %	ΔC_{2h} , %	ΔC_{12h} , %	ΔC_{74h} , %
40	50	20,0	45,3	55,0	11,2	43,4	33,1
80	50	24,2	47,2	48,3	29,2	53,6	36,4
40	500	73,2	84,7	86,4	55,0	58,0	60,0
80	500	63,2	83,2	86,3	55,2	60,7	61,2

Table 2. Concentration change of cobalt chloride and copper sulfate solutions.

The results of PGU sorptive properties research in static conditions correspond with the research results obtained under dynamic conditions. Ability of synthesized powdered PGU materials to sorb metal ions in dynamic conditions was examined for $CoCl_2$ and $CuSO_4$ water solutions of various concentration. PGU60 sorbent was used. Concentration of initial and final solutions was controlled using electron spectroscopy in Vis-region by the change of band intensity responding to d-d transitions in aqua ion $[Co(H_2O)_6]^{2+}$ and d-d transitions in aquaion $[Cu(H_2O)_6]^{2+}$ (near 510 nm and 810 nm respectively). Calibration graphs were plotted for 0,05; 0,075; 0,1; 0,15, 0,2 M concentrations of $CoCl_2 \times 6H_2O$ and $CuSO_4 \times 5H_2O$. Experiment was carried out at 18 °C. Solutions of copper and cobalt salts with initial concentrations of 0,1 M and 0,05 M were pumped through the column filled with the sorbent powder. m_{PGU}=7 g; $V_{solution}$=50 cm³. Under dynamic conditions the time of interaction ranged from 60 to 120 seconds. Fig. 7 illustrates the character of Co^{2+} and Cu^{2+} ion sorption.

After dynamic contact of copper salt with sorbent during 60 seconds absorbance of filtrate (D) falls to 1,08 in comparison with the initial solution absorbance value of 1,4. It responds to 40% decreasing of copper ions in solution. For 0,1M solution of cobalt chloride after dynamic contact with PGU40 there is observed absorbance fall from 0,52 to 0,48 that responds to extracting of 20% metal ions. For 0,05M solution of cobalt chloride is achieved 40% decreasing of cobalt ion concentration after 60 sec contact. Calculated concentration of cobalt ions in final solutions was 0,08M for initial 0,1M and 0,03M for initial 0,05M. The dynamic sorption of metal ions with PGU40 sorbent is more effective for solutions with lower concentration.

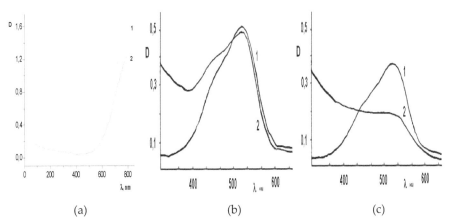

(a)   (b)   (c)

Figure 7. The electron spectra of water solutions of metal salts before (1) and after (2) passing through the column filled with powdered PGU40

According to (Bergmann et al., 2008) the mechanism of ion capture by water soluble polysaccharides mainly related to formation of complexes of chelate structure. Research of complex formation of PGU with metal ions allows conclude of chelate structure of formed complexes. In particular chelate structure of copper ion (2+) complexes with PGU is confirmed in (Hubina et al., 2010). Using results of (Bergman et al., 2008; Hubina et al., 2010) we can assume that cobalt ions also form chelate structures with PGU.

Analysis influence of cross-linking degree of PGU on complexing metal ions with functional groups of PGU demonstrated that variation of hydroxyl and urethane groups balance in the system allows to achieve effective control over holding metal ions in PGU matrix.

Peculiarities of "PGU-copper ion" complex formation were studied using electron paramagnetic resonance method (EPR). Copper ions were introduced into PGU matrix via pumping of 0,1M water solution of CuSO4 through column filled with PGU60 (the way it described above for dynamic sorption conditions). Then metal-containing PGU sample was dried and the EPR spectra of bivalent copper were recorded. Fig. 8 demonstrates EPR spectrum of dried PGU60 after interaction with copper sulfate water solution.

Recorded spectra are characterized with anisotropy of g-factor and appearance of hyperfine structure in the region of $g_\parallel$ that is usually concerned to tetragonal chelate complexes of bivalent copper. Hyperfine structure components are broadened as a result of superposition of signals from tetragonal copper complexes which can differ both in symmetry distortion and in nearest chemical surrounding. Integral intensity of obtained EPR spectra depends on the balance of hydroxyl and urethane group in PGU while electron spin parameters are almost unchanged. That points on preferable interaction of metal cation with one of the components of PGU. This conclusion corresponds with the regularities of bivalent copper ion complexing with mono saccharides.

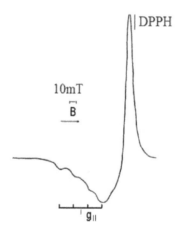

Figure 8. The representative EPR spectrum of copper containing PGU

Experimental analysis of influence of metal ions on water molecules self diffusion process in swelled polysaccharide gel was carried out by the method of quasi-elastic neutron scattering.

For description of experimental dependences there was calculated the values of general coefficient of water self diffusion in swelled gel $D = D_{singl} + D_{col}$ as well as values of its D_{singl} (single particle) component and D_{col} (collective) component. (Fig 9b).

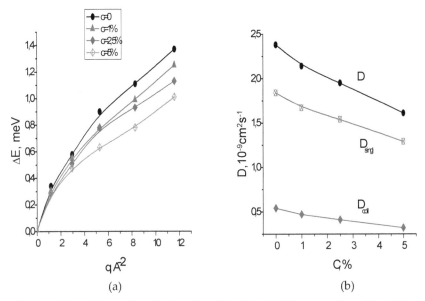

Figure 9. Dependence of energy broadening of quasi-elastic peak ΔE *vs* squared transferred kinetic moment q for various polymer–metal salt ratio (a) and concentration dependencies of D, D_{singl} and D_{col} (b).

Obtained data demonstrate that addition of metal salt to water and increasing of solution concentration decreases the coefficient of water molecules self diffusion in swelled xanthan gel both for general coefficient and for its components.

That indicates the swelled polymer density growth in presence of metal ions. Such condensation effect can be explained taking into account ability of transition metal chelates to form charge-transfer complexes with electron-donor centers of polymer resulting in creation of additional 'coordination juncs' both in swelled linear polysaccharide and in chemically cross-linked PGU.

Thus, polyglucanurethanes based on water soluble polysaccharide xanthan and blocked polyisocyanate are able to extract heavy metal ions from their salt water solutions. Variation of hydroxyl and urethane groups balance at the cross-linking stage provides effective regulation of complexing and capturing of metal ions with polyglucaneurethanes.

5. Biodestruction

Used plastics utilization had become an important problem. One of the ways of polymer and in particular of polyurethane waste reclamation is biodegradation of plastic. The ability to biodegradation was analized on the PGU exposed in the medium of common soil microorganisms association as well as into the medium of resistive microbial association isolated from soils polluted by chlorine-organic pesticides. (Hubina et al., 2009). There were explored biodegradable properties of two types of PGU20: based on microbial polysaccharide xanthan (PGU20) and methylcellulose (PGU20-cellulose).

The resistant microbial association with working name "Micros" was isolated from soil polluted by chlorine-organic pesticides. This association has high destructive activity with respect to chlorine-organic, organic-phosphorous, simm-triazine and other pesticide groups. It was supposed, that "Micros" is able to specialize to exotic substratum and could decompose polyglukanurethane systems due to utilization of this polymer as carbon or nitrogen source. To compare destruction of the PGU by common soil microorganisms the soil native-born microbial association was isolated from pollution-free chernozem soil. It has never contacted with pollutants in soil.

Changes in polymer structure after exposure in microbial medium were analyzed by FTIR spectroscopy, optical microscopy and thermogravimetry. Fig presents IR – spectra of PGU20 after contact with aggressive and natural microbial medium. Evaluation of redistribution of bonds in destructed PGU was provided by analysis of the location, width and intensity of valence vibrations band of C-O-C intercycle group of polysaccharide (805 cm^{-1}) and also by analysis of location and intensity of band amide II) in PGU urethane bridges (1600 ÷ 1660 cm^{-1}).

In PGU spectrum after the sample contact with aggressive medium (spectrum 3 fig 10) it is seen resolving of complex band at 1600 – 1660 cm-1 (N-H vibrations) comparing to initial PGU20 spectrum (Fig.10 spectrum 1). This may be caused by destruction of essential part of urethane bonds of PGU in aggressive medium.

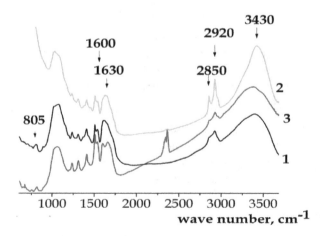

Figure 10. IR spectra of PGU20 based on xanthan 1 – initial PGU, 2 – PGU after contact with natural microbial medium, 3 – PGU after contact with aggressive microbial medium

Other type of PGU destruction is observed in natural medium (fig. 10 Spectrum 2). It is approved by redistribution of intensities of valence vibrations of C-O-C groups comparing with initial PGU and by absence of prominent changes in area of amide II. The band at 805 cm-1 responding to intercycle C-O-C bonds disappears. It is accompanied by increasing of intensity of the band responding to free OH-group (3430 cm-1) and C-H bonds (2850 and 2920 cm-1).

Another evidence of sufficient urethane bonds destruction in PGU20 in aggressive medium is seen on microphoto of PGU films after aggressive destruction (Fig. 11). There can be distinguished fibrous polymer elements, usual for fibrous of initial polysaccharide xanthan, which formed after urethane destruction. This effect is absent on the micro photos of PGU sample after destruction in natural microbial medium .

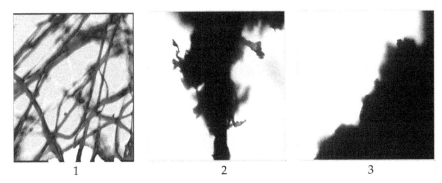

1 – initial polysaccharide fibers; 2 - effect of aggressive microbial medium; 3 – effect of natural microbial medium.

Figure 11. Micro images of PGU films after contact with natural and aggressive microbial mediums

IR spectra of PGU20-cellulose present the difference between this polymer biodestruction and biodestruction of PGU20-xanthan (Fig. 12)

In particular, the intensity of the band in area 1600 – 1660 cm-1 falls both for natural medium and aggressive one. (Fig.13 stectra 2 and 3). Intensity of 809 cm-1 band decreases (valence vibrations of intercycle bonds C-O-C) after exposure of PGU20-cellulose in natural microbial medium (Fig.12 spectrum 2) differing from intensity of this band in PGU20-cellulose spectrum after contact with aggressive medium. Bands corresponding to symmetric and asymmetric vibrations of saccharide groups C-O-C are nearly unchanged.

Efficient difference in biodestruction character of PGU20-xanthan and PGU20-cellulose allows conclude that xanthan based PGU is destructed mainly in its saccharide side-chains (containing D-mannose, D- glucuronic acid and D-glucose).

IR data correlate with results of TGA analysis. Table 3 shows difference in character of TG, DTG and DTA curves for PGU20-cellulose after contact with aggressive and natural medium. On DTG curves of PGU20-cellulose after natural and aggressive medium it is seen that stage of thermal destruction in temperature interval 200 – 300 oC is moved into district of higher temperatures while in PGU40-xanthan this stage is splitted and differs for natural and aggressive medium.

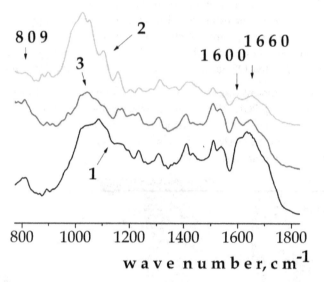

Figure 12. IR spectra PGU20-cellulose 1 – initial PGU20-cellulose; 2 – PGU after contact with natural microbial medium, 3 – PGU after contact with aggressive microbial medium.

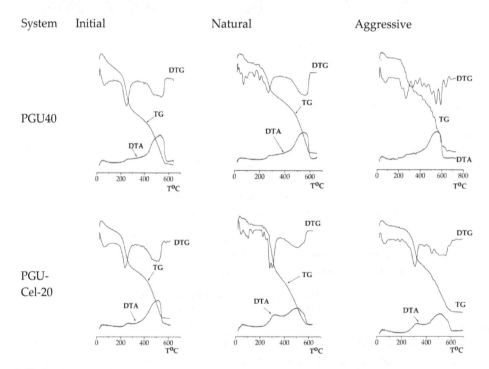

Table 3. IR spectra of PGU40 and PGU-cellulose-20 after biodestruction

6. Conclusions

New hydroxylcontaining polyurethane networks (polyglucanurethane) with various urethane group content were obtained based on microbial exopolysaccharide xanthan and latent blocked polyisocyanate using environment friendly technique. Obtained polyglukaneurethanes possess advanced thermal and chemical (both alkali and acid) resistance. Thermal resistance of PGU grows with increasing of urethane group amount. Were as with growth of urethane group content in the system the amount of absorbed water declines.

Study of properties of polyglucanuretanes reveals their ability to quantitative extraction of phenol and bivalent metal ions from water solution, while controlling sorption properties of the material with polysaccharide cross-linking degree. According to electron spectroscopy and EPR data the main mechanism of ion capture by polyglucanurethane consists in formation of "polymer-metal ion" chelate complexes.

Biodestruction research demonstrate that polyglucanurethane degradation in aggressive microbial media occurs via polysaccharide fragmentation due to urethane bonds cleavage. On the contrary microbial association that was isolated from pollution-free chernozem disintegrate the polysaccharide chains predominantly.

But both in aggressive microbial association and in natural microbial association that was isolated from pollution-free soil the direct relation was observed between destruction process intensity and percentage of polysaccharide OH-group substitution in PGU. The lower percentage of OH-groups was substituted, the more intensive destruction process was observed.

Author details

Nataly Kozak and Anastasyia Hubina
Institute of Macromolecular Chemistry National Academy of Sciences of Ukraine, Ukraine

Acknowledgement

The authors acknowledge Prof. Gvozdiak R.I, Dr. Dankevich L. and Dr. Vocelko S. (Institute of microbiology and virology of NAS of Ukraine) for help in the study of polyglukanurethane biodegradation in various microbial environments and for providing of various producents microbial exopolysaccharides.

7. References

Bergmann D., Furth G., Mayer Ch.Binding of bivalent cations by xanthan in aqueous solution. *International Journal of Biological Macromolecules.* 2008, Oct 1;43(3):245-51, 0141-8130

Crini G. Recent developments in polysaccharide-based materials used as absorbents in wastewater treatment. *Progress in Polymer Science.* 2005. V. 30, pp. 38 – 70, 0079-6700

Gvozdyak R.I., M.S. Matyshevskaya, Y.F. Grigoriev, O.A. Litvinchuk. (1989). *Microbial polysaccharide xanthan*, Naukova Dumka, 5120009670, Kyiv.

Hubina A.V., Kozak N.V., Nizelskii Yu.M. Hydroxyl containing polyurethane networks based on Xanthan and blocked polyisocianates and their interactin with phenol water solutions. *Polymer Journal*. 2009.-№1, pp. 58 – 61, 0203-327

Hubina A., Dankevich L., Kozak N., Yamborko N. 2009. Biodegradable Microbial Exopolysaccharide Based Polyurethane Networks for Phenols Sorption from Water Solutions. Odessa. 2009

Hubina A., Klepko V., Kozak N., Vasilkevich V., Slisenko V. Interaction of cross-linked polyglucanurethanes with transition metal salts water solutions. *Physics and Chemistry of Polymers (Tver)*. 2010. V. 16, pp. 214 - 219

Kozak N.V., Nizelskii Y.M. Polymer composition modification with blocked polyisocyanates. *Chemistry and chemical technology problems*, 2002.- №3, 0321-4095

State Standart (Ukraine) 12020-72

Chapter 8

Seed Oil Based Polyurethanes: An Insight

Eram Sharmin, Fahmina Zafar and Sharif Ahmad

Additional information is available at the end of the chapter

1. Introduction

Seed oils [SO] are cost-effective, eco-friendly and biodegradable in nature. They bear functional groups such as carboxyls, esters, double bonds, active methylenes, hydroxyls, oxirane rings and others, amenable to several derivatization reactions. Their abundant availability, non-toxicity and rich chemistry has established SO as focal point of polymer production, e.g., production of polyesters, alkyds, epoxies, polyols, polyethers, polyesteramides, polyurethanes and others. The escalating prices of petro-based chemicals, environmental and health concerns have further beckoned the enhanced utilization of SO as polymer precursors. SO have attracted enormous attention as potential source of platform chemicals, at both laboratory and industrial scale. Today, oil-seed bearing crop plants are being raised and modified for uses in areas covering biodiesel, lubricants, folk medicines, cosmetics, plastics, coatings and paints.

SO based polyurethanes [PU] occupy an important position due to their simple preparation methods, outstanding properties and versatile applications in foams, coatings, adhesives, sealants, elastomers and others [1-4]. In general, PU are prepared by chemical reaction of a diol, polyol or any oligomer or polymer containing hydroxyl groups, with an aliphatic, cycloaliphatic or aromatic isocyanate. SO serve as green precursors to diols, polyols and other –OH containing polymers offering greener raw materials in PU synthesis, replacing their petro-based counterparts. The choice of SO polyol or isocyanate is governed by the end-use application of SO PU ranging from soft and flexible to rigid PU foams, thermoplastic to thermosetting PU, flexible films to tough, scratch-resistant, impact resistant coatings and paints. It is well known that the structure of a triglyceride is very complex. Every SO has a characteristic fatty acid profile. Amongst the same species, the composition of triglycerides in a particular SO varies by weather conditions of crops and soil. Triglycerides vary by their fatty acid chain lengths, presence as well as the position of double bonds and degree of unsaturation of the constituent fatty acids. The structures of natural SO and their derivatives, i.e., epoxies, diols, polyols, polyesters and alkyds are complex. Thus, the properties of PU obtained from SO derivatives depend on a number of factors such as (i) the type, composition and distribution of fatty acid residues in the constituent triglycerides, (ii) the number, distribution, site of hydroxyls (in the middle or closer to the end of the triglyceride chain) and level of unsaturation in the fatty triester chains of the parent diol or polyol, (iii) the type, position and structure of isocyanates used and (iv) the urethane content of the final PU [5-8].

The ingredients for the preparation of SO based PU generally comprise of a diol or polyol (containing active hydrogens) derived from SO and an isocyanate, aliphatic and aromatic such as 1,6-hexamethylene diisocyanate [HMDI], isophorone diisocyanate [IPDI], cyclohexyl diisocyanate [CHDI], L-Lysine Diisocyanate [LDI], toluylene 2,4-diisocyanate or toluylene 2,6-diisocyanate [TDI], 4,4'- methylenediphenyl diisocyanate [MDI], naphthylene 1,5-diisocyanate [ND]. PU are prepared by polyaddition reaction between the two components, often in presence of a chain extender, cross-linker or a catalyst. Today, several environment friendly routes for the production of PU have cropped up. Research has been focussed on the preparation of PU from non-isocyanate routes, and also on the use of fatty acid based isocyanates for PU production, considering the persisting challenges of polymer industry and immediate attention sought towards environmentally benign chemicals and chemical routes for the same [9-18].

SO based PU are generally flexible in nature. Generally, these PU show low Tg due to the presence of long aliphatic chains and also poor mechanical properties (owing to the presence of dangling chains). The thermal stability of SO based PU is also lower since the decomposition of urethane bonds starts at 150-200°C. Javni et al have studied the decomposition of PU from Olive, Peanut, Canola, Corn, Soybean, Sunflower, Safflower and Castor oils [7]. The decomposition involves the dissociation of urethane bonds to isocyanate, alcohol, amine, olefin and carbon dioxide. The properties of PU depend upon their crosslinking density as well as chemical composition. In an execellent review, Petrovic has highlighted the different properties of PU prepared from polyols obtained by different methods. As the properties of polyols depend upon the methods of preparation, so also the properties of PU derived therefrom. He has presented a brief outline of the effect of polyols prepared by epoxidation, hydroformylation, ozonolysis, effect of crosslinking density, and type of isocyanate on the properties of PU. He described the effect of the structure of polyols prepared by epoxidation followed by ring opening with methanol, HCl, HBr, and by hydrogenation of epoxidized Soybean oil, and showed that PU obtained from these polyols showed relatively higher glass transition temperatures and improved mechanical properties. Halogenated polyols obtained via ring opening by HCl and HBr gave PU that were less stable than ones without halogens, and had higher Tg (70°C –80°C) than the latter. Polyols obtained via hydroformylation crystallize below room temperature while those derived through hydrogenation reveal crystallization at temperature higher than room temperature. PU from non-halogenated polyols had higher thermal stability than brominated (100°C) or chlorinated polyols (160°C). Polyols with primary hydroxyls give more stable PU than their counterparts with secondary hydroxyls. PU with high crosslink density show higher thermal stability. Hydrolytic stability of PU also depends on the degree of crosslinking, temperature, and physical state of PU. In SO based PU, although SO have ester groups susceptible to hydrolysis, long hydrophobic fatty acid chains cause shielding effect. Under highly humid conditions, urethane bonds undergo hydrolysis producing amine and carbon dioxide [1, 19].

SO are rich in various functional groups such as double bonds, active methylenes, esters, hydroxyls and others. These undergo several chemical transformations yielding numerous derivatives with versatile applications as inks, plasticizers, lubricants, adhesives, coatings and paints. Amongst various SO derivatives, those containing hydroxyl groups are used in the production of PU. These include SO based diols, triols, polyols, polyesters, alkyd, polyesteramide, polyetheramide and others (Figure 1).

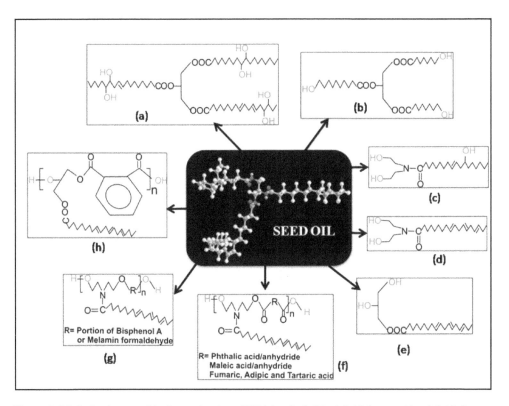

Figure 1. SO derivatives used in the production of PU (a) polyol, (b) triol, (c) fatty amide triol, (d) fatty amide diol, (e) monoglyceride, (f) polyesteramide, (g) polyetheramide, and (h) alkyd.

The chemistry of SO based PU is very vast, governed by several factors such as the type of diol or polyol, type of isocyanate, preparation method of diols or polyols, type of chain extender, cross-linker, reaction temperature and other reaction conditions. In this chapter we have focussed on the preparation, structure and properties of PU obtained from diols, triols and polyols derived by amidation of SO termed as "SO alkanolamides". In the proceeding sections, we have also discussed the modifications of the said SO alkanolamides based PU at the forefront of PU chemistry such as SO based metal containing PU, PU hybrids, composites for applications mainly in coatings, paints and foams.

2. SO based diols

The most excessively used SO based diol in PU production is fatty amide diol or fatty alkan-diol-amide [FAD] (Figure 2). FAD is obtained by the base catalysed amidation of SO with diethanolamine. The structure of FAD is determined by the fatty acid composition of the starting SO.

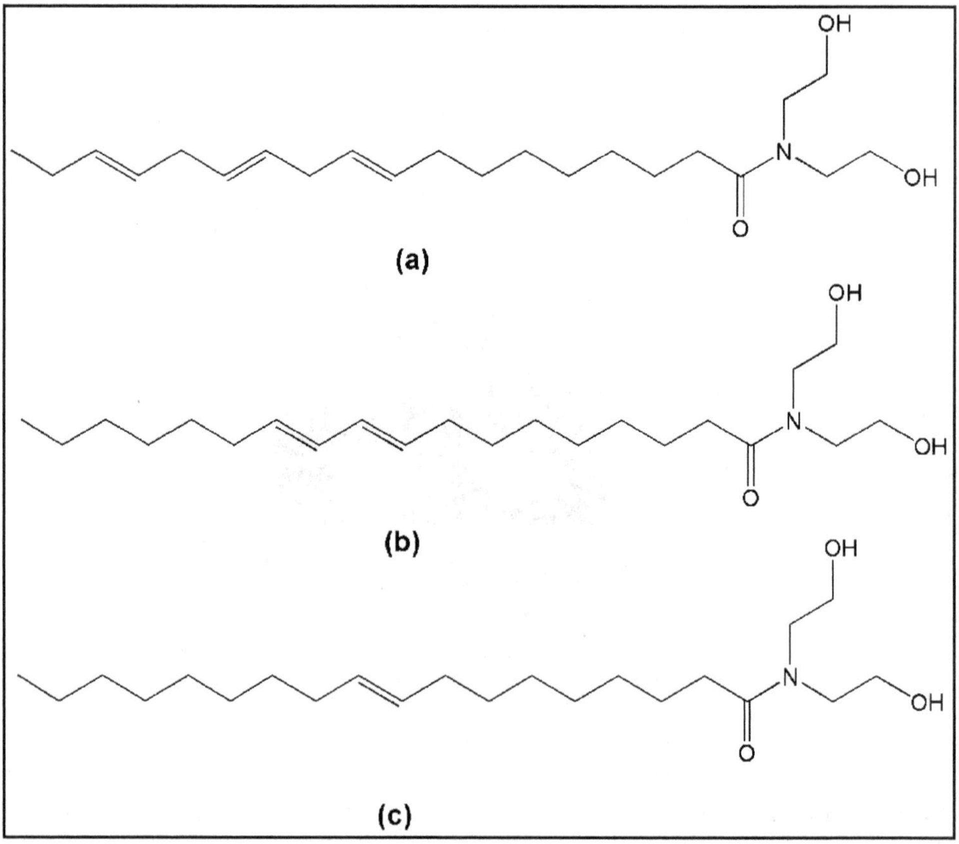

Figure 2. Figure 2. FAD from (a) Linseed (35.0-60.0 % Linolenic acid), (b) Soybean (43.0-56.0 % Linoleic acid) and (c) Karanj (44.5–71.3 % oleic acid), Nahor (55-66% oleic acid), Jatropha (37-63 % oleic), Olive (65-80 % oleic acid)

The reaction proceeds by nucleophilic acyl substitution bimolecular (SN2) reaction mechanism. As the name suggests, FAD contains an amide group, with two hydroxyl ethyl groups directly attached to amide nitrogen along with the pendant aliphatic chain of SO. FAD is derived from different SO such as Linseed (*Linum ussitassimum*), Soybean (*Glycine max*), Karanj (*Pongamia glabra*), Nahor (*Mesua ferrea*), Jatropha (*Jatropha Curcus*), Olive (*Olea europea*), Coconut (*Cocos nucifera*) and others [20-30] (Figure 2). FAD is used as raw material for various polymers such as PU, poly (esteramide) and poly (ether amide).

3. PU from SO FAD

FAD can be treated with an isocyanate such as TDI, IPDI, HMDI, MDI, ND, CHDI and LDI forming poly (urethane fatty amide) (Figure 3) [FADU] [31].

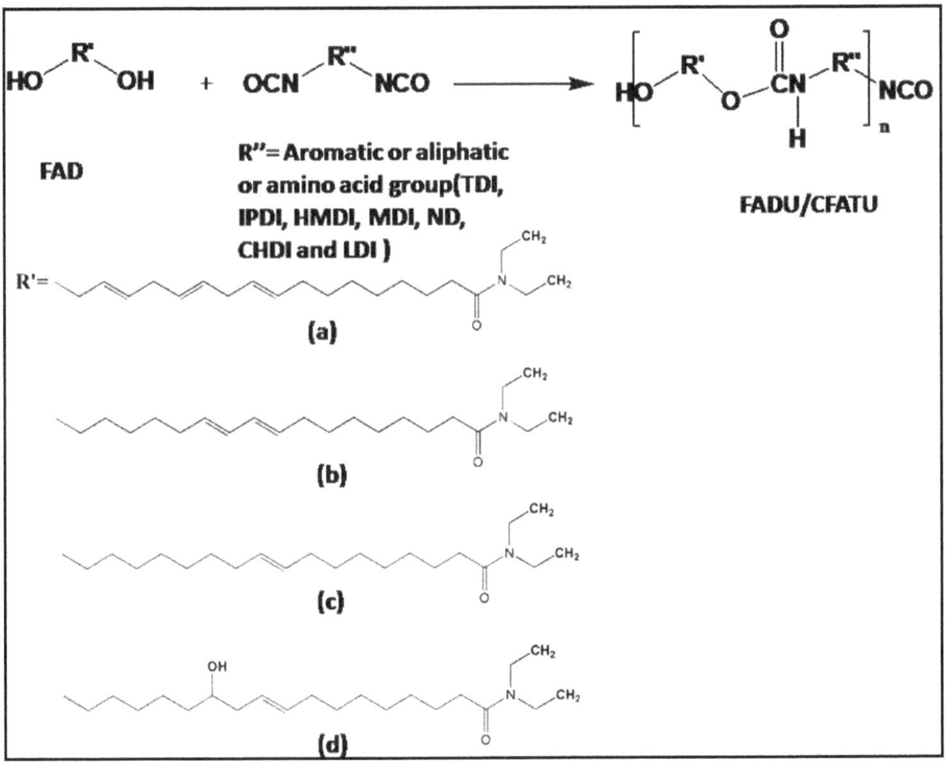

Figure 3. FADU from (a) Linseed, (b) Soybean (c) Karanj, Nahor, Jatropha, Olive and (d) Castor

For the first time, Linseed oil [LO] derived FAD [LFAD] was treated with TDI by one-shot technique to prepare PU [LFADU] as introduced by Ahmad et al [32] (Figure 4).

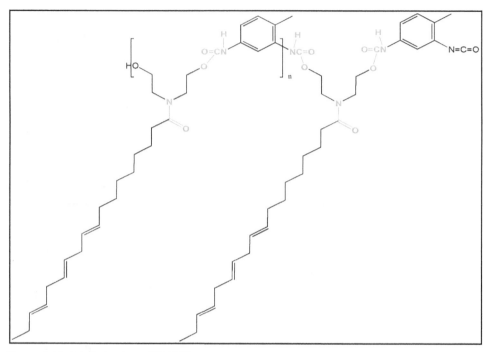

Figure 4. Chemical structure of LFADU.

The structure of LFADU was established by spectral analyses. FTIR, ^{1}H NMR and ^{13}C NMR spectra showed the presence of the main characteristic absorption bands of parent SO [32]. Along with these bands, additional absorption bands are observed supporting the presence of urethane groups in the backbone of LFADU such as those at 3375cm^{-1} for hydroxyl groups, 1716.11 cm^{-1} for urethane carbonyl (str), 1227.56cm^{-1} for C–N of urethane groups, 1735cm^{-1} typical for carbonyl (str) of TDI. The characteristic peaks for hydrogen bonded and non-hydrogen bonded protons of –HNCOO– appear at 7.99–7.82ppm and 7.1–6.9ppm, respectively. The aromatic ring protons of TDI occur at 7.5–7.22 ppm. The peaks of – HNCOOCH$_2$– are observed at 4.1–3.9ppm and for CH$_3$ of TDI appear at 2.25ppm. ^{13}C NMR spectrum also shows the presence of characteristic peaks of LFADU at 17ppm (CH$_3$ of TDI), 143.97ppm {–NH–(C O)–O–} and 137.46, 136.2, 134.4, 125.94, 125.4, 116.0 ppm (aromatic ring carbons of TDI). TGA thermogram of LFADU has shown four step degradation pattern, at 260 °C (27% weight loss), 360 °C (21% weight loss), 505 °C (40% weight loss), 640 °C (9% weight loss) corresponding to the degradation of urethane, ester, amide and hydrocarbon chains, respectively.

PU from Karanj or *Pongamia glabra* [PGO] oil [PFADU] has also been prepared by similar method. PU obtained from both FAD showed similar structure except for the difference in the structure of pendant fatty amide chains attributed to the variation in the structure of the parent SO chain [33] (Figure 5).

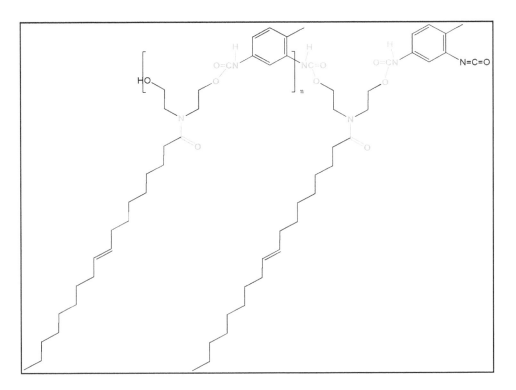

Figure 5. Chemical structure of PFADU or NFADU

The thermal degradation of PFADU was observed at 177°C and 357°C, with 5% weight loss occurring at 200°C attributed to the entrapped solvent and moisture, 10wt% loss at 225 °C, attributed to the decomposition of urethane moieties, 50wt% and 80wt% losses at 390 °C and 455 °C, respectively, attributed to the degradation of the aromatic ring and aliphatic pendant fatty alkyl chains, respectively.

It was observed that in both LFADU and PFADU, beyond 1.5moles loading of TDI, formation of some lumpy aggregates occurred. Upto 1.5 moles of TDI addition, it is speculated that the isocyanate groups of TDI react with hydroxyl groups of FAD forming PU linkages. Beyond this amount, any additional isocyanate added reacts with the urethane groups of LFADU or PFADU forming allophanate groups (secondary reaction). The final PU attains very high viscosity and crosslinking, so much so that the formation of lumpy aggregates occurs and PU is deprived off its free flowing tendency.

Karak and Dutta have reported the production of PU by amidation and urethanation of methyl ester of *M. Ferrea* or Nahor oil [NO], rich mainly in triglycerides of linoleic, oleic, palmitic and stearic acids. They investigated the structure and physico-chemical aspects of FADU from NO [NFADU] [26].

3.1. PU as coatings

LFADU has free –OH, –NCO, aliphatic hydrocarbon chains (from parent LO), amide and urethane groups, which make it an excellent candidate for application in protective coatings (Figure 4). LFADU coatings undergo curing at ambient temperature (28-30°C) by three stage curing phenomenon, (i) solvent evaporation (physical process), (ii) reaction of free –NCO groups of LFADU with atmospheric moisture, and (iii) auto-oxidation. These coatings show good scratch hardness (2.5kg), impact resistance (200lb/inch), bending ability (1/8inch) and chemical resistance to acid and alkali. PU from PGO [PFADU] has shown moderate antibacterial behavior against *Salmonella* sp. with good scratch hardness (1.9kg), impact resistance (150lb/inch), bending ability (1/8inch), and gloss (46 at 45°) [33]. LFADU coatings have shown superior coating properties than PFADU owing to the fatty acid composition of parent oils (PGO, a non-drying oil has higher content of oleic acid while LO, a drying oil, is rich in linolenic acid).

Karak and Dutta have reported the use of NFADU coatings with very good alkali resistance (Figure 5)[32].

3.2. PU as hybrids

Organic-inorganic hybrid materials have been developed with FADU as organic constituent and metal/metalloid as inorganic component to improve the performance and broaden the applications of PU (Figure 6).

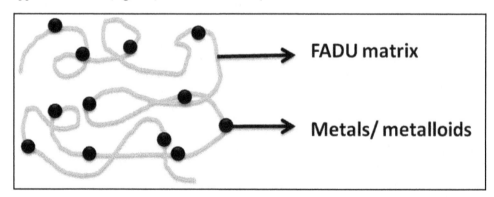

Figure 6. PU as hybrids

In one report, Zafar et al. have prepared organic-inorganic hybrids by using boric acid as inorganic content and PFADU as organic matrix [B-PFADU] [34]. B-PFADU was characterized by standard spectral techniques and physico-chemical methods. B-PFADU performed well as protective coatings in terms of physico-mechanical and chemical resistance tests. B-PFADU showed no change in water and xylene upto 15 days. However, slight deterioration in performance was observed in alkali and acid media, correlated to the presence of -O-B-O- which is susceptible to hydrolysis on exposure to these media. B-

PFADU showed high activity against *E. coli* (Zone of inhibition: 21-30 mm) and moderate activity against *S. aureus* (Zone of inhibition: 16-20 mm). The reason can be the presence of urethane, amide, and hydroxyl groups in the polymer backbone, which can presumably interact with the surface of microbes, causing antibacterial action. B-PFADU can be used as an antibacterial agent as well as coating material.

In another work, Ahmad and co-workers developed LFADU hybrid material with tetraethoxy orthosilane [TEOS] as inorganic constituent in the hybrid material [Si LFADU] by in situ silylation of LFAD with TEOS (at 80 °C) followed by urethanation with TDI (at room temperature) [35]. Along with the typical absorption bands for LFADU, additional absorption bands were observed at 484 cm^{-1} (Si–O–Si bending), 795 cm^{-1} (Si–O–Si sym str) and 1088 cm^{-1} (Si–O–Si assym str) in FTIR due to the presence of -Si–O–Si- bond in the hybrid backbone. Hydroxyl value decreases while refractive index and specific gravity increase with the loading of TEOS in Si LFADU, supporting the formation of the hybrid materials by insitu siylation and urethanation reaction. Optical micrograph of Si LFADU showed the presence of SiO$_2$ particles surrounded by LFADU (Figure 7).

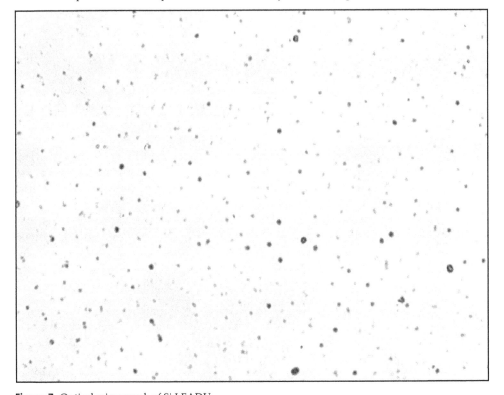

Figure 7. Optical micrograph of Si LFADU

Si LFADU formed hybrid coatings by simple curing route at ambient temperature, over mild steel panels of standard sizes with improved gloss and scratch hardness. SiO2 domains also improved adhesion with the penal surface exhibiting good scratch hardness, bending ability (1/8 inch) and impact tests (150 lb/inch) correlated to the synergism showed by both the components, LFADU backbone imparting flexibility and gloss, while the inorganic domains conferring excellent adhesion and hardness [36].

The corrosion rate (CR) of Si LFADU is much lower (3.08 × 10^{-4} mm per year) relative to LFADU (3.124 mm/year) In 3.5wt% HCl, with inhibition efficiency (IE%) 99.77. In 3.5% NaOH, CR and IE% were found as 1.26 × 10^{-3} mm per year and 99.34, respectively. Si LFADU formed uniform and well adhered coating over the metal substrate which prohibits the permeation of corrosive media. The protection mechanism is purely through barrier action attributed to the hydrophobic inorganic content [37, 38]. Coating remained intact when subjected to corrosive media for 192 h as supported by the constant value of polarization resistance (Rp = 1.22 × 104 Ohm in NaOH and 7.7 × 105 Ohm for HCl). Thermal studies showed four step degradation, thermal stability increasing with higher inorganic content, with two glass transition temperatures (Tg) as observed at 115 °C and 155 °C in DSC thermogram with safe usage upto 200 °C.

3.3. PU as composites

Composite materials from FADU have not been prepared yet. In their recent research, Zafar et al have developed composites from FADU using metal oxides and organo-montmorillonite clay [OMMT] (Cloisite 30B; modified by an alkyl ammonium cation bearing two primary hydroxyl functions, alkyl group is tallow containing ≈65% C18, ≈30% C16, and ≈5% C14, Southern Clay product) as modifiers added in very lower amounts to FADU matrix (unpublished work). The sole aim behind the development of these composites was the improvement in performance of FADU materials in terms of thermal stability, physico-mechanical and corrosion or chemical resistance performance, and also antimicrobial behavior relative to the pristine material for high performance applications. MMT occurred as nano-sized aggregates with size ranging from 37 to 100 nm as observed by Transmission Electron Micrography [TEM]. The thermal stability of FADU/ MMT was found to increase with increased MMT loading. These composites may be used as protective coatings in future. Zafar et al have also developed FADU/MnO composites, with good antifungal behavior. MnO occurred as needles self-assembled in "lemon slices" morphology as observed in optical micrograph (Figure 8). The said composite material may find application as antimicrobial agent in coatings and paints.

PU composites were prepared by the dispersion of conducting polymer poly (1-naphthylamine) [PNA] in LFADU matrix in amount ranging from 0.5–2.5 wt% by conventional solution method as reported by Riaz et al [39]. At lower levels, the composites were very fragile in nature. The polymerization of PNA occurred through N–C(5) linkages of 1-naphthylamine units as confirmed by FTIR spectra. PNA also reacted with free isocyanate groups of TDI forming urea linkages, as also supported by spectral analysis. UV visible spectra also confirmed the formation of urea linkages between LFADU and PNA. TEM micrographs showed the average particle size as 17–27 nm. Nanoparticles appeared as smaller aggregates which later on formed larger aggregates. XRD analysis revealed purely amorphous nature of composites. With the increase in the loading of PNA in the composites, the distortion and torsional strain increased in the composites due to higher urea linkages. It was found that as the percent loading of PNA in the composites increased, their electrical conductivity values also increased; however, these values fell in the semi-conducting range, which was much higher relative to the conductivity values obtained with very high loading of PNA in previously reported composites. The improved electrical conductivity values of LFADU/PNA composites can be correlated to the hydrogen bonding and urea type linkages formed between the two polymers, which provide the path to charge conduction [40,41].

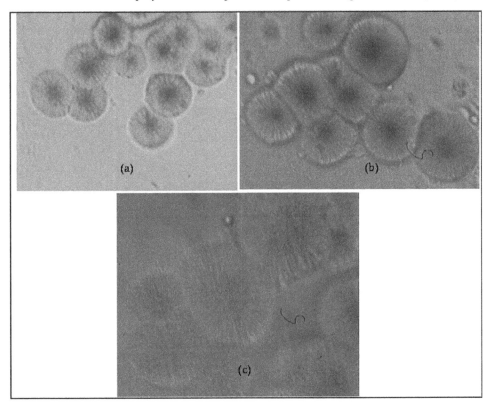

Figure 8. Optical micrographs of FADU/MnO (a) 100 X, (b) 200 X, (c) 500 X

4. SO based triol

Castor oil (CO) is obtained from seeds of *Ricinus communis* or Castor belonging to the family *Euphorbiaceae*. It is non edible oil. The crop is cultivated around the world because of the commercial importance of its oil. India is the world leader in castor production and dominates the international CO trade. Worldwide castor production was about 1.4 million metric tons during the year 2009 with an average yield of about 956 kg ha^{-1}. Ricinoleic acid

(12- hydroxy -9- octadecenoic acid), hydroxyl containing fatty acid, is the major component of CO, and constitutes about 89% of the total fatty acid composition. Castor seed products have widespread applications in many industries like paints, lubricants, pharmaceuticals and textiles. Today, the importance of these products has grown manifolds because of their biodegradable and eco-friendly nature.

Due to the inherent hydroxyl functionality, CO stands as an excellent natural raw material for the development of PU. CO derived PU are flexible due to long aliphatic triglyceride dangling chains [42]. CO has three hydroxyl moieties in one triglyceride molecule. On amidation, CO yields fatty amide triol or alkan-triol-amide [FAT], which bears two hydroxyl ethyl groups directly attached to amide nitrogen, as well as one hydroxyl group in the pendant fatty chain obtained from the parent CO. Thus, CO derived FAT [CFAT] acts as SO derived triol (Figure 9). Rao et al prepared acrylated CFAT as a multifunctional amide for photocuring [31, 43, 44].

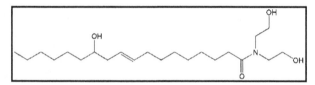

Figure 9. CO (87-90 % Ricinoleic acid) derived FAT [CFAT]

5. PU from SO FAT

CFAT on chemical reaction with TDI by one shot technique yields CFATU (Figure 10), with an additional crosslinking site (hydroxyl group) conferred by parent CO containing 89% ricinoleic acid. Contrary to LFADU and PFADU, where the permissible limit of TDI addition is 1.5 moles, in CFATU, at 1.2.moles of TDI addition, CFATU becomes highly viscous. The difference prevails due to the additional hydroxyl functionality in CFATU, which presumably makes possible higher inter and intra crosslinking sites relative to LFADU and PFADU. As also observed in LFADU and PFADU, the physico-chemical characteristics such as specific gravity, inherent viscosity and refractive index increase, while hydroxyl and iodine values decrease on increasing the content of TDI in PU. The thermal degradation occurred in the temperature range of 150–390 °C. The decomposition observed at earlier temperature range may be attributed to the thermal degradation of urethane linkages followed by the volatilization of the decomposition products while that at higher temperatures is correlated to the degradation of amide bond, aromatic ring and aliphatic alkyl chain of the fatty acid, respectively, followed by the volatilization of the decomposition products [45, 46].

5.1. PU as coatings

CFATU coatings have been prepared and their physico-mechanical, thermal as well as corrosion resistance behavior has also been investigated [31]. CFATU have been further modified for improvement in their performance as discussed in proceeding sections.

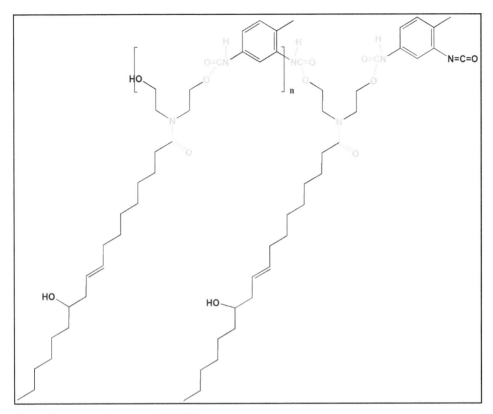

Figure 10. Chemical structure of CFATU

5.2. PU as hybrids

Ahmad et al have reported the preparation and characterization of metal containing CFATU [47]. They treated CFAT with zinc acetate (5, 10 and 15 wt%) and TDI (25–75 wt%) to prepare metal containing CFATU [MCFADU] "in situ" by microwave [MW] assisted preparation method in a domestic MW oven. During the preparation, it was observed that MCFATU with > 5wt% zinc acetate formed gel. While the conventional preparation method of LFADU, CFATU, PFADU and NFADU requires 8-12 hours, CFATU is obtained in 4 minutes by MW assisted preparation method. By molecular interactions with the electromagnetic field and heat generated by molecular collision and friction, the reaction is facilitated to occur in much reduced time periods under the influence of MW irradiations. In this reaction, hydroxyl groups of CFAT react with zinc acetate and TDI successively. Curing of MCFATU is a two step process occurring by solvent evaporation (physical phenomenon) followed by the second stage curing (a chemical process) where free –NCO groups of MCFATU react with the atmospheric moisture to form urethane and amino groups through addition reaction. MCFATU acted as good corrosion protective coatings for mild steel. The best physico-mechanical properties (scratch hardness 3.5 kg, impact resistance

150lb/inch, and bending ability 1/8 inch) were achieved at 5wt% loading of zinc acetate and 55 wt% of TDI, when evaluated by standard methods and techniques. The corrosion protection efficiency of the same was evaluated by potentiodynamic polarisation measurements [PDP] in aqueous solution of 3.5wt% HCl (CR 4.51 × 10-3 mm/year; IE% 96.23), 3.5wt% NaOH (CR 1.36 × 10-3 mm/year; IE% 90.81); 3.5wt%NaCl (CR 2.25 × 10-3 mm/year; IE% 94.50) and tap water (Cl- ion 63 mg/L; Conductivity 0.953 mS/A) (CR 3.56 × 10-3 mm/year; IE% 98.35).

5.3. PU as composites

CFATU composites were developed by the introduction of MMT clay [48] and nano sized ZnO by Zafar et al [unpublished work]. Morphology of the composites as studied by TEM revealed the presence of nanosized globules of size ranging between 15-120 nm in CFATU/OMMT (Figure 11), and ZnO in CFATU/ZnO composites occurring as nano-sized spindles of diameter 50-60nm (Figure 12). Both the composite materials may find application as corrosion protective coatings for mild steel. CFATU/ZnO composites also act as excellent antifungal agents against common fungal strains such as *Candida albicans*, *Candida krusei*, *Candida glabrata* and *Candida tropicalis*.

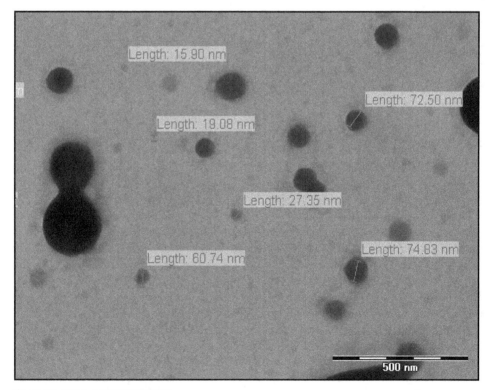

Figure 11. TEM of CFATU/OMMT

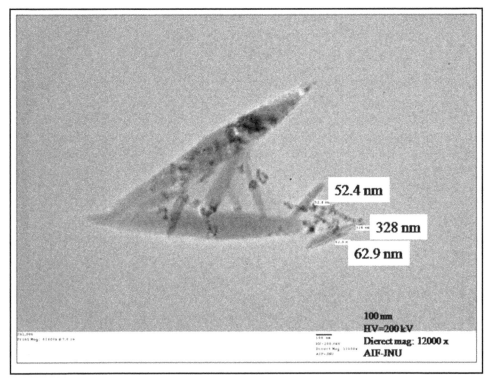

Figure 12. TEM of CFATU/ZnO composite

6. SO based polyols

SO derived polyols serve as the most important oleochemicals for PU production. Some of the SO polyols are derived through various chemical reactions such as epoxidation followed by hydration, ring opening by methanol, acids such as HBr, HCl, hydroformylation, ozonolysis of SO, others are naturally available polyols such as *Ricinus communis* or CO and Lesquerella oil. As discussed previously, the properties of PU also depend upon the type of polyol and the method by which the polyol is derived. In SO polyols, in general, hydroxyl groups are present in the middle of the triglyceride chains. Due to this, in cross linked polyols, the pendant or dangling chains provide steric hindrance to cross-linking, they do not support stress under load and act as plasticizers. In SO polyols, there is difference in the length of elastically active network chains (EANC) and elastically inactive network chains or dangling chains (DC). This variation is also passed onto their respective PU. The properties of PU thus also depend upon the content of EANC and DC, i.e., the number and position of hydroxyl groups. The number of hydroxyls on each chain in SO polyols and their stereochemistry are also variable depending upon the hydroxylating agents, hydroxylating method and other reaction conditions. For example, polyols obtained by ozonolysis and hydrogenation bear hydroxyl groups at terminal position.

SO polyols may also undergo amidation with diethanolamine to yield fatty amide polyols (FAP) [49]. Similar to FAD and FAT, FAP also house two hydroxyl ethyl amides directly attached to amide nitrogen and multiple hydroxyl groups located on the pendant fatty alkyl chains, which were part of the triglyceride molecule in parent SO bearing double bonds at the site of hydroxyl groups in FAP. The properties of SO polyols obtained by either method mentioned previously, also influence the properties of FAP. Hydroxylated, hydroxymethylated, carboxylated SO followed by their amidation yield polyols with higher number of hydroxyl groups with improved distribution [50-56]. These are ideal candidates to produce PU foams. The approach has been accomplished on CO, LO, PO, Rapeseed, Safflower, Soybean oils and refined bleached deodorized Palm Kernel Olein. These polyols have been used as non-ionic surfactants in the household and cosmetic industries and also to produce PU foams on treatment with suitable isocyanates. Such polyols prove to be advantageous over CO as they can be incorporated in higher amounts during PU formulations. Foams obtained show improvement in terms of high close cell contents, good dimensional stability and compression strength.

6.1. PU foams from SO FAP

Alkanolamide polyols serve as excellent starting materials for PU foams [50-56]. The variations in fatty acid components of starting SO, number and position of hydroxyl groups and also the presence of dangling chains in the polyol confer differences in performance and cellular structures in PU foams. The hydroxyl content of PU determines the suitability of PU foams ranging from flexible to rigid foams. A. Palaniswamy et al. produced PU foams from FAP derived from PO and Polymeric Diphenylmethane Diisocyanate (PMDI) by hand foaming. It was found that the decreased FAP content led to increase in compressive strength and density of PU foam [50]. In another research work, they have produced PU foams from PMDI and CO, in the presence of stannous octoate as catalyst and Tegostab by hand mixing process with carbon dioxide as the blowing agent generated from reaction between excess PMDI and water. PU foams with varying FAP content, catalyst and molecular weight of poly propylene glycol were studied with respect to their effect on density and compression strength [50, 51].

6.2. PU foams from SO based FAD epoxies

The epoxidized oil based alkanolamides are also classified as polyols for PU foams [56, 57]. PU show low thermal stability, thus with view to improve the thermal stability and mechanical properties of PU, heterocyclic groups such as isocyanurate, imide, phosphazene and oxazolidone, are incorporated in SO PU. The latter is formed by the chemical reaction between an oxirane ring and isocyanate in presence of a catalyst, the approach improving both thermal stability as well as stress-strain properties of the modified product with respect to the pristine material. PU foams derived from epoxidised alkanolamides show better compression strength, thermal conductivity, close cell contents and dimensional stability relative to plain alkanolamide PU. However, in some examples, a part of epoxy content is lost during amidation reaction occurring at higher temperatures. Thus, it became imperative to determine ideal reaction conditions for amidation to retain maximum number of epoxidized rings, which was attempted by Lee et al and characterized by high performance liquid chromatography and gas chromatography [55, 56].

7. SO based polymers for PU production

As discussed earlier, SO undergo numerous transformations yielding various derivatives. Some of these derivatives bearing (inter or intra located) hydroxyls serve as excellent starting materials for PU production. A large number of PU are prepared from SO polymers such as polyesters, alkyds, polyesteramides, polyetheramides [Figure 1], which find profound applications in paints and coatings.

8. Summary

FAD, FAT and FAP serve as good starting materials for PU production. LFADU, PFADU, CFATU have similar structural characteristics; the difference being due to the pendant fatty amide chains attributed to the fatty acid composition of the parent SO. LFADU, PFADU,and CFATU are formed at a particular NCO/OH ratio. An astonishingly abnormal rise in viscosity was observed in LFADU and PFADU above 1.5moles and in CFATU above 1.2moles of addition of TDI, followed by the formation of lumpy aggregates. According to the general chemistry of PU, a particular NCO/OH ratio is required for a particular application. The best properties in PU are achieved when this ratio is kept as or closer to 1 or 1.1, i.e., when one equivalent weight of isocyanate reacts with one equivalent weight of polyol, to achieve the highest molecular weight. In certain applications this ratio is kept well below the stoichiometry (higher hydroxyl content relative to isocyanate) to obtain low molecular weight PU for applications as adhesives and coatings as described in the chapter. NCO/OH ratio is varied by the formulator based on the type of end use application of PU. Properties of PU mentioned here also depend on the chemical route of raw materials (polyol, isocyanate), functionality and type of the raw materials (diol, triol, polyol and isocyanates-aliphatic, aromatic), the number of urethane groups per unit volume, non-isocyanate PU, as well as other structural differences such as the presence of modifiers (acrylics, metals, nanosized metal oxides, MMT clay).

The preparation through MW technique offers advantages of reduced times and improved yield. Most of these PU are used for coatings and foams. The incorporation of inorganic constituent led to improved thermal and hydrolytic stability as well as coating performance of PU. Another area that is presently being explored is the preparation of green PU from fatty isocyanates or non-isocyanate PU. Due to their numerous applications and advantages SO PU have been extensively studied and extensive research is still going on.

Author details

Eram Sharmin and Fahmina Zafar*
Department of Chemistry, Jamia Millia Islamia (A Central University), New Delhi, India

Sharif Ahmad
Materials Research Lab, Department of Chemistry,
Jamia Millia Islamia (A Central University), New Delhi, India

Acknowledgement

Dr Fahmina Zafar (Pool Officer) and Dr.Eram Sharmin (Pool Officer) acknowledge Council of Scientific and Industrial Research, New Delhi, India for Senior Research Associateships against grant nos. 13(8385-A)/2010-POOL and 13(8464-A)/2011-10 POOL, respectively. They are also thankful to the Head, Department of Chemistry, Jamia Millia Islamia(A Central University), for providing support to carry out the work.

9. References

[1] Petrović Z. S. Polyurethanes from vegetable oils. Polymer Reviews 2008; 48:109-155.

[2] Lligadas G., Ronda J.C., Galia`M., Cadiz V. Plant oils as platform chemicals for polyurethane synthesis:current state-of-the-art. Biomacromolecules 2010; 11: 2825-2835.

[3] Desroches M., Escouvois M., Auvergne R.,Caillol S., Boutevin B. From vegetable oils to polyurethanes: synthetic routes to polyols and main industrial products. Polymer Reviews 2012; 52 (1): 38-79.

[4] Pfister D.P., Xia Y., Larock R.C. Recent advances in vegetable oil-based polyurethanes. Chem Sus Chem 2011; 4(6):703-17.

[5] Zlatanic A., Petrovic Z. S., Dusek K. Structure and properties of triolein-based polyurethane networks. Biomacromolecules 2002; 3 (5): 1048-1056.

[6] Guo A., Cho Y., Petrovic Z. S. Structure and properties of halogenated and nonhalogenated soy-based polyols. J Polym Sci Part A: Polym Chem. 2000; 38 (21): 3900-3910.

[7] Javni I., Petrovic Z. S., Guo A., Fuller R. Thermal stability of polyurethanes based on vegetable oils. Journal of Applied Polymer Science 2000; 77 (8): 1723-1734.

[8] Ligadas G., Ronda J. C., Galia M., Cadiz V. Novel silicon-containing polyurethanes from vegetable oils as renewable resources. Synthesis and properties. Biomacromolecules 2006; 7 (8): 2420-2426.

[9] Bähr M., Mülhaupt R. Linseed and soybean oil-based polyurethanes prepared via the non-isocyanate route and catalytic carbon dioxide conversion. Green Chemistry 2012;14: 483-489.

---

* Corresponding Author

[10] Guan J., Song Y., Lin Y.,Yin X., Zuo M., Zhao Y., Tao X., Zheng Q. Progress in study of non-isocyanate polyurethane. Industrial Engineering Chemistry Research 2011; 50: 6517-6527.

[11] Gonzalez-Paz R.J., Lluch C., Lligadas G., Ronda R.C., Galia M., Cadiz V. A Green approach toward oleic and undecylenic acid-derived polyurethanes. Journal of Polymer Science Part A. Polymer Chemistry 2011; 49: 2407-2416.

[12] Parzuchowski P.G., Jurczyk-Kowalska M., Ryszkowska J., Rokicki G. epoxy resin modified with soybean oil containing cyclic carbonate groups. Journal of Applied Polymer Science 2006; 102: 2904-2914.

[13] Javni I., Hong D.P., Petrovic Z.S. Soy-based polyurethanes by nonisocyanate route. Journal of Applied Polymer Science 2008; 108: 3867-3875.

[14] Hojabri L., Kong X., Narine S.S. Fatty acid-derived diisocyanate and biobased polyurethane produced from vegetable oil: synthesis, polymerization, and characterization. Biomacromolecules 2009; 10 (4): 884-891.

[15] Hojabri L., Kong X., Narine S.S. Biomacromolecules 2010; 11: 911-918

[16] Palaskar D.V., Boyer A., Cloutet E., Alfos C., Cramail H. Synthesis of biobased polyurethane from oleic and ricinoleic acids as the renewable resources via the AB-type self-condensation approach. Biomacromolecules 2010; 11: 1202-1211.

[17] Tamami B., Sohn S., Wilkes G.L. Incorporation of carbon dioxide into soybean oil and subsequent preparation and studies of nonisocyanate polyurethane networks. Journal of Applied Polymer Science 2004; 92: 883-891.

[18] Matsumura S., Soeda Y., Toshima K. Perspectives for synthesis and production of polyurethanes and related polymers by enzymes directed toward green and sustainable chemistry. Appllied Microbiology Biotechnology 2006; 70: 12–20.

[19] Gast, L.E., Schneider W.J., Mc Manis G.E., Cowan, J.C. Polyesteramides from linseed and soybean oils for protective coatings: Diisocyanate-modified polymers. *Journal of the American Oil Chemists' Society* 1969; 46 (7): 360-364.

[20] Gast L.E., Schneider W.J., Cowan, J.C. Polyesteramides from linseed oil for protective coatings. *Journal of the American Oil Chemists' Society* 1966; 43(6): 418-421.

[21] Gast, L.E., Schneider W.J., Cowan, J.C. Polyesteramides from linseed oil for protective coatings low acid- value polymers. *Journal of the American Oil Chemists' Society* 1968; 45(7): 534-536.

[22] Ahmad S., Ashraf S M., Yadav S., Hasnat A. A polyesteramide from Pongamia glabra oil for biologically safe anticorrosive coating. Progress in Organic Coatings 2003; 47 (2): 95-102.

[23] Zafar F., Sharmin E., Ashraf S. M., Ahmad S. Studies on poly (styrene-*co*-maleic anhydride)-modified polyesteramide-based anticorrosive coatings synthesized from a sustainable resource. Journal of Applied Polymer Science 2004; 92: 2538-2544.

[24] Zafar F., Ashraf S.M., Ahmad S. Studies on zinc-containing linseed oil based polyesteramide. Reactive & Functional Polymers 2007; 67: 928-935.

[25] Raval D.A., Patel V.M., Parikh D.N. Streptomycin release from N,N-bis(2-hydroxyethyl) fattyamide modified polymeric coating. Reactive and Functional Polymers 2006; 66 (3): 315-321.

[26] Dutta S., Karak N., Synthesis, characterization of poly (urethane amide) resins from Nahar seed oil for surface coating applications. Progress in Organic Coatings 2005; 53: 147-152.

[27] Khan N.U., Bharathi N. P., Shreaz S., Hashmi A.A. Development of water-borne green polymer used as a potential nano drug vehicle and its in vitro release studies. Journal of Polymers and the Environment 2011; 19 (3): 607-614.

[28] Bharathi N. P., Khan N. U., Alam M., Shreaz S., Hashmi, A. A. Edible oil-based metal-containing bioactive polymers: synthesis, characterization, physicochemical and biological studies. Jouranl of Inorganic and Organometallic Polymers and Materials 2010; 20:839–846.

[29] Alam M, Alandis N.M., Microwave Assisted Synthesis Of Urethane Modified Polyesteramide Coatings From Jatropha Seed Oil. Journal of Polymers and the Environment 2011; 19 (3): 784-792.

[30] Alam M, Alandis N.M., Microwave assisted synthesis and characterization of olive oil based polyetheramide as anticorrosive polymeric coatings (communicated).

[31] Kashif M. Development and characterization of poly (urethane-amide) protective coating materials from renewable resource. Thesis submitted to Jamia Millia Islamia (A Central University), New Delhi, India.

[32] Yadav S., Zafar F., Hasnat A., Ahmad S., Poly (urethane fatty amide) resin from linseed oil—A renewable resource. Progress in Organic Coatings 2009; 64 (1): 27-32.

[33] Kashif M., Zafar F., Ahmad S., Pongamia glabra seed oil based poly(urethane–fatty amide). Journal of Applied Polymer Science 2010; 117: 1245–1251.

[34] Zafar F., Kashif M., Sharmin E., Ahmad S. Studies on boron containing poly(urethane fattyamide). Macromolecular Symposis 2010; 290: 79-84.

[35] Ahmad S., Zafar F., Sharmin E., Garg N., Kashif M. Synthesis and characterization of corrosion protective polyurethanefattyamide/silica hybrid coating material. Progress in Organic Coatings 2012; 73 (1): 112-117.

[36] Ooij W.J.V., Zhu D., Stacy M., Mugada T., Gandhi J., Puomi P., Corrosion protection properties of organofunctional silanes—an overview. Tsinghua Science & Technology 2005;10 (6): 639-664.

[37] Phanasgaonkar A., Raja V.S., Influence of curing temperature, silica nanoparticles- and cerium on surface morphology and corrosion behaviour of hybrid silane coatings on mild steel. Surface and Coatings Technology 2009; 203(16): 2260-2271.

[38] A.S. Vuc, M. Fir, R. Jese, A. Vilcnik, B. Orel, Structural studies of sol–gel urea/polydimethylsiloxane barrier coatings and improvement of their corrosion inhibition by addition of various alkoxysilanes. Progress in Organic Coatings 2008; 63 (1): 123-132.

[39] Ashraf S.M., Ahmad S., Riaz U., Development of novel conducting composites of linseed-oil-based poly(urethane amide) with nanostructured poly(1-naphthylamine). Polymer International 2007; 56:1173-1181.

[40] Rodriguesa P.C., Akcelrud L. Networks and blends of polyaniline and polyurethane: correlations between composition and thermal, dynamic mechanical and electrical properties. Polymer 2003; 44 (22):6891-6899.

[41] Chiang L.Y., Wang L.Y., Kuo C.S., Lin J.G. and Huang C.Y. Synthesis of novel conducting elastomers as polyaniline-interpenetrated networks of fullerenol-polyurethanes, Synthetic Metals 1997; 84 (1-3):721-724 .

[42] Mutlu H., Meir M.A.R. Castor oil as a renewable resource for the chemical industry. European Journal of Lipid Science and Technology 2012; 112 (1): 10-30.

[43] Rao B.S., Palanisamy A. Synthesis, photo curing and viscoelastic properties of triacrylate compositions based on ricinoleic acid amide derived from castor oil. Progress in Organic Coatings 2008; 63: 416-423.

[44] Rao B.S., Palanisamy A. Photo-DSC and dynamic mechanical studies on UV curable compositions containing diacrylate of ricinoleic acid amide derived from castor oil. Progress in Organic Coatings 2007; 60:161-169.

[45] Somani K, Kansara S, Parmar R, Patel N. High solid polyurethane coatings from castor oil based polyester polyols. International Journal of Polymer Materials 2004; 53:283-293.

[46] Szycher M (1999) Szycher's Handbook of polyurethane, 2nd edn.CRC Press, Sterling Biomedical, Inc, Lynnfield MA, Michael Szycher, Cardio-Tech International, Woburn, Massachusetts.

[47] Zafar F., Mir M.H., Kashif M., Sharmin E., Ahmad S. Microwave assisted synthesis of bio based metallopolyurethaneamide. Journal of Inorganic and Organometallic Polymers and Materials 2011; 21 (1): 61-68.

[48] Zafar F., Sharmin E., Zafar H., Ahmad S. Synthesis and characterization of bio-nanocomposites based on polyurethanefattyamide/ organo-montmorillonite. 2011; communicated

[49] Sharmin E., Akram D., Ahmad S. Polyol from linseed oil for waterborne coatings: synthesis and characterization. International conference. Polymer Science & Technology: Vision & Scenario (APA-2009) at New Delhi, India on Dec. 17-20, 2009

[50] Palanisamy A, Karuna M. S. L., Satyavani T., Rohini Kumar D. B., Development and Characterization of Water-Blown Polyurethane Foams from Diethanolamides of Karanja Oil. Journal of the American Oil Chemists' Society 2011; 88 (4): 541-549.

[51] Palanisamy A, Rao B. S., Mehazabeen S., Diethanolamides of castor oil as polyols for the development of water-blown polyurethane foam. Journal of Polymers and the Environment 2011; 19:698-705.

[52] Khoe T.H., Otey F., Frankel E.N., Cowan J.C. Polyurethane foams from hydroxymethylated fatty diethanolamides. Journal of the American Oil Chemists' Society 1973; 50:331-333.

[53] Khoe T.H., Frankel E.N. Rigid polyurethane foams from diethanolamides of carboxylated oils and fatty acids. Journal of the American Oil Chemists' Society 1976; 53:17-19.

[54] Shapiro SH (1968) Commercial nitrogen derivatives of fatty acids. In: Pattison ES (ed) Fatty acids, their industrial applications. Marcel Dekker, New York, pp 77-154

[55] Lyon C.K., Garret V.H., Frankel E.N. Rigid urethane foams from hydroxymethylated castor oil, safflower oil, oleic safflower oil, and polyol esters of castor acids. Journal of the American Oil Chemists' Society 1974; 51:331-334.

[56] Lee C.S., Ooi T.L., Chuah C.H., Ahmad S. Synthesis of palm oil-based diethanolamides. Journal of the American Oil Chemists' Society 2007; 84:945-952.

[57] Badri K.H., Othman Z., Ahmad S.H. Rigid polyurethane foams from oil palm resources. Journal of Materials Science 2004; 39:5541-5542.

Chapter 9

Biobased Polyurethane from Palm Kernel Oil-Based Polyol

Khairiah Haji Badri

Additional information is available at the end of the chapter

1. Introduction

Polyurethanes are block copolymers containing segments of low molecular weight polyester or polyether bonded to a urethane group (-NHCO-O). Traditionally, these polymers are prepared by reacting three basic materials; polyisocyanates, hydroxyl-containing polymers (polyester or polyether polyol) and chain extender, normally low molecular weight diol or diamine (such as 1, 4-butanediol or 1, 4-dibutylamine).

Polyols are generally manufactured by one or two possible chemical routes, namely alkoxylation and esterification. Alkoxylation, by far is the most common route, involves the reaction between a hydroxyl or an amine-containing initiator (such as sucrose, glycerol) and either propylene- or ethylene oxide. A molecular weight of up to 6000 can be obtained by extending the polymer chain with the addition of alkylene oxide. This product is suitable for more flexible polyurethanes in cushioning and elastomeric applications. The alkylene oxide used in this process is derived from mineral oil via the petroleum industry. Propylene for instance, is derived from the petroleum cracking process and is then converted to propylene oxide before being further converted to polyol by reaction with an amine or hydroxyl-containing initiator such as glycerol.

At present, most polyols used in polyurethane industry are petroleum-based where crude oil and coal are used as starting raw materials. However, these materials have been escalating in price and rate of depletion is high as well as required high technology processing system. This necessitates a look at utilizing plants that can serve as alternative feed stocks of monomers for the polymer industry. Moreover, with increasing annual consumption of polyurethane, its industrial waste is a serious matter. In Europe and the United States of America for instance, government regulations encouraged recycling of materials to avoid excessive usage of landfill area. However, with thermosetting behavior of polyurethane the recycling activity is difficult and limited. The best alternative is biodegradation. Biodegradable polymers have widely been used in pharmaceutical industry such as suture usage, wound-dressings, surgical implants and medicine delivering system. But there are still some usage limitations either due to high production cost or its low performance. This performance can be achieved by chemical and physical modification of these materials through combination of biodegradable and non-biodegradable materials.

Polyurethane based on polyester has been known to be more biodegradable than from polyether. Utilization of renewable resources to replace petrochemicals in polyurethane industry has attracted attention of many technologists. Most of these renewable resources are forest products. Palmeri oil, vernonia oil, castor oil and cardanol oil (extracted from the cashew nut shell) have been used to synthesize polyurethane polyols with multiple functionality to replace the petrochemical-based polyols (Pourjavadi et al. 1998 and Bhunia et al. 1998). Castor oil has long been used in the polyurethane industry. Relatively, it is stable to hydrolysis due to its long fatty acid chain but sensitive to oxidation due to the present of unsaturated fatty acid. Commercially, it can only be used in the coating and adhesive industries.

Polyester polyols are generally consisted of adipic acid, phthalic anhydride, dimer acid (dimerized linoleic acid), monomeric glycol and triol. It has low acid number (normally 1-4 mg KOH/g) and low moisture content (less than 0.1%). These properties are not easily achieved unless a high-technology processing method is applied. Due to these industrials requirements, polyester polyols are usually supplied at higher price compared to polyether polyols. Polyether polyols on the other hand, are commercially produced from catalytic reaction of alkylene oxide i.e.propylene oxide or ethylene oxide to di- or polyfunctional alcohol. Its functionality is four and above and is useful in the production of rigid foam. It can also be produced with the presence of di- or polyfunctional amine i.e. diethanolamine when high reactivity is required (such as laminated continuous panel production). Important properties specified in polyurethane industry for polyols are as summarized in Table 1.

Classification	Flexible foam / Elastomer	Rigid / Structural foam
Molecular Weight	1,000 to 6,500	400 to 1,200
Hydroxyl value, mgKOH/g	28 to 160	250 to 1,000
Functionality	2.0 to 3.0	3.0 to 8.0

Table 1. Technical requirements for polyols used in polyurethane industry (Wood 1990).

The lower the equivalent weight of polyol is, the higher the rigidity of the polyurethane. These contributed to higher compressive strength, modulus, thermal stability and dimensional stability polyurethanes. If the equivalent weight is excessively low, the resulting polymer becomes more friable and required more isocyanate especially for the production of rigid polyurethane foam (Berlin and Zhitinkina 1982).

Natural occurring oils and fats are water-insoluble substances originated from vegetable, land or marine animal known as triglycerides. A triglyceride is the reaction product of one molecule of glycerol with three molecules of fatty acids to yield three molecules of water and one molecule of a triglyceride. The molecular weight of the glycerol portion (C_3H_5) of a triglyceride molecule is 41. The combined molecular weight of the fatty acid radicals (RCOO-) varies. Natural oils can undergo a number of chemical reactions such as hydrolysis, esterification, interesterification, saponification, hydrogenation, alkoxylation, halogenation, hydroxylation, Diels-Alder reaction and reaction with formaldehydes. Polyester is a high molecular weight chemical with ester group –O-C=O- as repeating unit. It is achieved by polycondensation and esterification of carboxylic acid with hydroxyl-containing compounds.

Lauric oil or better known as lauric acid is the main source of fatty acids. The only lauric oils available to the world market are coconut oil and palm kernel oil. The oil palm is a monocotyledon belonging to the Elaeis Guiness species. Palm kernel oil (PKO) is obtained from the kernel part of the oil palm fruit. The percentage of unsaturated fatty acids is much lower compared to palm oil as shown in Table 2. PKO consist of 80 percent saturated fatty acid and 10% of each polyunsaturated and unsaturated fatty acid. Palm oil on the other hand, consist of 53% saturated fatty acid, 10% polyunsaturated and 37% unsaturated fatty acids. The higher the unsaturated fatty acid contents the unstable it is when exposed to heat. The reactivity increases substantially if the double bond are conjugated (separated by one single bond) or methylene-interupted (separated by a –CH_2 unit). PKO contains only traces of carotene.

Vegetable Oil	Saturated Fatty Acid, %							Unsaturated Fatty Acid, %					
								Enoic				Dienoic	Trienoic
Carbon Chain	C8	C10	C12	C14	C16	C18	>C18	<C16	C16	C18	>C18	C18	C18
Palm Oil				1-6	32-47	1-6				40-52		2-11	
Palm Kernel Oil	2-4	3-7	45-52	14-19	6-9	1-3	1-2		0-1	10-18		1-2	

Table 2. Fatty acid contents in palm oil and palm kernel oil (Khairiah Haji Badri 2002).

Two major reactions occurred during polymerization of polyurethane. First, the reaction of isocyanate with water yields a disubstituted urea and generates carbon dioxide. This is called the blowing reaction because the carbon dioxide is acting as an auxiliary-blowing agent. The second reaction is between the polyfunctional alcohol (polyol) and the isocyanate (Fig. 1).

R-N=C=O + R'-O-H   →   R-NH-C(O)-O-R
Isocyanate   Polyol   Polyurethane
R-N=C=O + H-O-H   →   R-NH-C(O)-O-R' + R-N-H + CO_2
Isocyanate   Water   Polyurethane
R-N=C=O + R-N-H   →   R-N-C-N-R
Isocyanate

Figure 1. Addition polymerization of polyurethane

It generates a urethane linkage and this is referred to as the gelation reaction. The isocyanate reacts slowly with alcohols, water and the unstable amino products without the present of catalyst. However, for most commercial requirements the acceleration of these reactions is required.

One characteristic of amorphous polymeric systems is the glass transition temperature, T_g that defines the point where the polymer undergoes a change from glassy to rubbery behavior. Considerable attention has been devoted over the last several years to these studies: synthesis of polyurethane polyol from PKO and the production of oil palm empty fruit bunch fiber-filled PU composites (Badri et al. 1999, 2000[a], 2000[b], 2001; Khairiah Haji Badri 2002; Badri et al. 2004[a], 2004[b]; Badri et al. 2005; Badri & Mat Amin 2006; Badri et al. 2006[a], 2006[b]; Mat Amin et al. 2007, Norzali et al. 2011[a], 2011[b]; Liow et al.; Wong & Badri 2010, Badri & Redhwan 2010;). These include intensive evaluation on the chemical, mechanical, thermal and environmental stress on the synthesized polyol and PU foam by looking at various scopes:

• Synthesis of the palm kernel oil-based polyol from refined, bleached and deodorized (RBD) palm kernel oil via esterification and polycondensation.
• Preparation of the polyurethane foam from the RBD PKO-based polyol and evaluation of its chemical, mechanical and thermal decomposition and glass transition temperature of the foam.

2. Vegetable oil-based polyurethane polyol

Several reports have been published in producing polyurethane from vegetable oils and some of them have even been patented (Arnold 1983, Chittolini 1999 & Austin et al. 2000). Focus was given to utilization of mixture of vegetable oils in the polyurethane system and not as raw materials to produce the polyurethane. Vegetable oils that are frequently used are soybean oil, safflower oil, corn oil, sunflower seed oil, linseed oil, oititica, coconut oil, palm oil, cotton seed oil, peritta oil, olive oil, rape seed oil and nuts oil. Researches carried out using these oils were focusing on full usage of materials found abundance in certain area such as production of polyurethane foam from mixture of starch and triol polycaprolactone (Alfani et al. 1998) and mixture of starch, soybean oil and water *(Fantesk)* (Cunningham et al. 1997). Polyurethane products based on vegetable oils like nuts oil, soybean oil, corn oil, safflower oil, olive oil, canola oil and castor oil (Nayak et al. 1997, Bhunia et al. 1998, Mohapatra et al. 1998, Javni et al. 1999) exhibited high thermal stability. In Malaysia, the Malaysian Palm Oil Board (MPOB) has taken the initiative to produce polyol from the epoxidation and alcoholysis of palm oil (Ahmad et al. 1995, Siwayanan et al. 1999). An early finding has indicated that when natural oils or fats are epoxidized, they react with polyhydric alcohols to produce polyols. A study by Guthrie and Tait (2000) has successfully produced an ultraviolet (UV) curable coating from epoxidized and unprocessed palm oil, and epoxidised palm olein.

These researches however, are pointing to one direction that is synthesizing polyester. Polyester may be defined as heterochain macromolecules containing repeating ester groups (-

COO-) in the main chain of their skeletal structures. Most useful routes to polyester synthesis of carboxylic acids are step growth or direct polyesterification (condensation polymerization) and ring opening polymerization of lactones. The former is suitable for synthesis of aliphatic polyester where it utilizes primary and secondary glycols where the primary hydroxyl groups being esterified more readily. The removal of liberated water from the process is carried out by stirring and percolation of inert gas such as nitrogen, N_2. If a volatile monomer is used (i.e. glycol), an excess amount with respect to dicarboxylic acid (10 %w/w) should be added to compensate for losses caused by evaporation at high temperature. Side reactions may occur usually at 150°C and above which leads to changes in polymer structure and reduces molecular weight distribution of the polyester (Jedlinski 1992).

Esterification is one of many substitution reactions of carboxylic acids and their derivatives that involve tetrahedral addition intermediates. The extension of mechanism of carbonyl addition is as shown below. The best leaving group is the weakest base. In addition, reaction of ester with hydroxylamine (:NH$_2$OH) gives N-hydroxyamides (known as hydroxamic acids). This is the point where it is vital to add some reactivities to the existing polyester by addition of the amide group to form polyesteramide (Loudon 1988).

$$
\begin{array}{ccccccccc}
\text{O} & & & & \text{OH} & & \text{O} & & \\
\| & & & & | & & \| & & \\
\text{R-C-X} & + & \text{Y-H} & \rightleftharpoons & \text{R-C-X} & \longrightarrow & \text{R-C-Y} & + & \text{X-H} \\
& & & & | & & & & \\
& & & & \text{Y} & & & & \\
\end{array}
$$

leaving   addition   elimination
group

Polyurethanes are possible to decompose by prolonged contact with water, diluted acids or moist heat (causes swelling and slow hydrolysis, particularly in some ester-type polyurethanes), chlorine bleach solutions (may cause yellowing and decomposition) and prolonged exposure to light (discoloration of derivatives of aromatic isocyanates) (Roff et al. 1971). The dimensional stability of foams is a time-dependent property that receives considerable attention. Disregarding cold aging at -15±2°C, humid aging (70± 2°C at 95±5% relative humidity) is usually a prime property. Humid aging requirements (specifications) are determined by the end use of the foam. A foam that has expanded and the shrunk is considered, as a first approximation, to be caused by the effect of plasticization by heat and moisture that would allow the stresses built into the foam at the gel to relax, which will then allow the foam to return to a lower energy state. For urethane foams specifically, high thermal stability results in excellent dimensional stability over a large temperature range.

3. Green material and technology

The RBD palm kernel oil (viscosity of 65 cps, specific gravity of 0.99 g/ml, and moisture content of 0.02%) was obtained from Lee Oilmill Sdn Bhd, Kapar, Klang, Malaysia and was used as received without further purification. Polyhydric compounds consisted of dietanolamine, DEA (purity of 99.8%, hydroxyl value of 1057 mg KOH/g and functionality of 2, with viscosity of 236 cps and moisture content of 0.05%) and ethylene glycol, MEG (hydroxyl value of 1122 mg KOH/g) were supplied by Cosmopolyurethane (M) Sdn Bhd, Pelabuhan Klang, Malaysia with the inclusion of potassium acetate which was manufactured by Merck (M) Sdn Bhd, Shah Alam Malaysia. Chemicals used for the preparation of polyurethane foam were crude MDI (2,4-diphenylmethane diisocyanate), tetramethylhexanediamine (TMHDA) and pentamethyldiethyltriamine (PMDETA) (Cosmopolyurethane (M) Sdn Bhd, Port Klang, Malaysia) and silicon surfactant (Tegostab B8408, Th. Goldschmidth, Singapore). The blowing (foaming) agent used was tap water.

DEA, MEG and potassium acetate were mixed homogeneously with a ratio of 90:7:3 to form the polyhydric compound. A mixture of this polyhydric compound with RBD PKO at stochiometric ratio was continuously stirred in a 2-L glass reactor and was reacted separately at three different temperature ranges: 165-175°C, 175-185°C and 185-195°C, each for 30 minutes. The nitrogen gas was flushed into the system throughout the process. The reflux flask was connected to a condenser and a vacuum pump to withdraw the water and excess of reagent from the system. The progress of the reaction was monitored by sampling at intervals. The samples collected were then analyzed. At the end of the reaction, the polyol produced was kept in a sealed cap glass jar for further analysis. 140 g of crude MDI was poured into 100 g mixture of resin (Table 3 and Appendix A).

Composition	Part by weight, pbw
RBD PKO Polyol	100
Tegostab B8408	2
TMHDA	0.3
PMDETA	0.15
Water	4.5
Total pbw	106.95
Ratio of 100 parts to MDI	100:140

Table 3. Formulation of palm-based polyurethane foam system.

The mixture was agitated vigorously using a standard propeller at a speed of 200 rpm for 10 seconds at 20°C (Fig. 2).

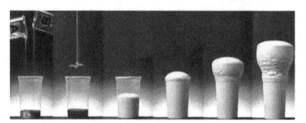

Figure 2. Polymerization of the palm-based polyurethane

The reaction time: cream time (CT), fiber/gel time (FT), tack-free time (TFT) and rise time (RT) was noted (Appendix B). The free-rise density (FRD) was calculated using equation (1).

$$\text{Free - rise density, FRD (kg / m}^3) = \frac{\text{(mass of foam and cup - mass of cup), kg}}{\text{capacity of cup, m}^3} \qquad (1)$$

The mixture was poured into a waxed mold, covered and screwed tight. The foam was demolded after 10 minutes. The molded density was determined using equation (2). The molded foam was conditioned for 16 hours at 23 ± 2°C before further characterization of the polyurethane foam.

$$\text{Molded density, MD (kg/m}^3) = \frac{\text{mass of molded foam, kg}}{\text{volume of molded foam, m}^3} \qquad (2)$$

Fourier Transform Infrared analysis of the RBD PKO polyol and palm-based PU was carried out on the Perkin Elmer Spectrum V-2000 spectrometer by Diamond Attenuation Total Reflectance (DATR) method. The samples collected during the intervals were scanned between 4000 and 600 cm^{-1} wavenumbers. For the former, two selected peaks (designated as peaks A and B) were used to monitor the progress of the reaction (derivatization).

Chromatography analyses were carried out on the former by thin layer chromatography followed by gas chromatography. A sample of 1 pph (part per hundred) by dilution in methanol was dropped on the silica plate with minimal diameter of about 0.5 mm and was applied 2 cm from the edge of the silica plate. The plate was removed once it traversed 2/3 of the length of the plate (normal length of a silica plate is 20 cm). The plate was placed in a chamber containing iodine crystals (iodine reacts with organic compounds to yield dark stain) after the methanol has all evaporated. The molecular weight was determined using gas chromatography coupled with mass spectrometer, GC-MS model Bruker 200 MHz with splitless inlet and HP5 (polar) column with flow rates of 1.0 μl/min. The oven was programmed to a temperature range of 100 to 280 °C at 6 °C/min.

Standard method ASTM D4274-88 (Standard Test Methods for Testing Polyurethane Raw Materials: Determination of Hydroxyl Numbers of Polyols) was used to determine the hydroxyl value of the polyol. The value calculated would be able to verify the FTIR peak ratio method for completion of derivatization process. The water content of the polyol was determined using the Karl Fischer Titrator model Metrohm KFT 701 series (ASTM D4672-00(2006) e1: Standard Test Methods for Polyurethane Raw Materials: Determination of Water Content of Polyols) while the viscosity of the polyol was determined using the Brookfield digital viscometer model DV-I (ASTM D4878-88: Standard Test Methods for Polyurethane Raw Materials- Determination of Viscosity of Polyols). The viscosity of the polyol is important in determining the flowability of the polyurethane resin in the foaming process where it is advantageous in the material consumption. The specific gravity was determined following ASTM D4669–07: Standard Test Method for Polyurethane Raw Materials: Determination of Specific Gravity of Polyols. Other physical characterizations were determination of cloud point, pH and solubility of polyol in methanol, benzene, acetone, ether and water.

The PU foams were characterized for their apparent molded and core densities, compression strength, dimensional stability and water absorption following standard method BS4370: Part 1:1988 (1996) Methods 1 to 5: Methods of test for rigid cellular materials. Foam samples were cut using into cubes of 100 mm × 100 mm × 100 mm in dimensions. A replicate of five specimens were used and carefully weighed using an analytical balance. The dimensions were measured following BS4370: Part 1:1988 (1996): Method 2. The apparent molded density was determined by using a simple mathematical equation, mass (kg)/volume (m^3). The core density is determined by the same method but using skinless foam. The compressive strength test was carried out on a Universal Testing Machine Model Testrometric Micro 350 following BS4370: Part 1:1988 (1996): Method 3 at 23 ±2°C. The specimens were cut into cubes of 50 mm × 50 mm × 50 mm in dimensions. The foam rise direction was marked and a crosshead speed of 50 mm/min was applied. The compression stress at 10% deflection, compression stress at 5% strain and compression modulus was noted. For the dimensional stability test, the specimens were cut into dimensions of 100 mm × 100 mm × 25 mm. The specimens were then put into a controlled temperature-humidity chamber each at –15 ± 2°C and 70 ± 2°C, 95 ± 5% relative humidity for 24 hours. Method 5A of BS4370: Part 1:1988 (1996) standard was followed. The specimens were remeasured and percentage of change in dimensions was calculated. These are then converted to percentage in volume change. The water sorption was carried out using method in Annex D BS6586: Part 1:1993. The specimens were cut into dimension of 50 mm × 50 mm × 50 mm.

The thermal decomposition of the polyurethane foam was measured using a thermogravimetric analyzer model Shimadzu TGA-50 with temperature ranging from room temperature to 600°C at heating rate of 10°C/min under nitrogen gas atmosphere. Samples were placed in alumina pan holders at a mass ranging from 5 to 15mg. The thermal property of the foam was determined using a Perkin Elmer Model DSC-7 differential scanning calorimeter interfaced to the Model 1020 Controller. The samples were analyzed from room temperature to 200°C at a heating rate of 10°C/min. Standard aluminum pans were used to analyze 10 mg samples under nitrogen gas atmosphere. The insulation value (k-factor or λ-value) of the polyurethane foam was determined using the Thermal Conductivity Analyzer model Anacon at testing temperature for cold plate at 25°C and hot plate at 35°C. The thickness of the specimens was 20-30 mm and method 7 of BS4370: Part 2: 1993 standard was followed.

The RBD PKO consists of triglycerides that when undergoes esterification form by products such as glycerol and other possible polyester network (Loudon 1988) as shown in Scheme 1 and Scheme 2. During the reaction, the acetate ion forms an intermediate, the carboxylic acids. These acids attack the lone pair in nitrogen atom in diethanolamine, DEA and formed the probable structure of the esteramide with hydroxyl terminal (Scheme 2)

R_1, R_2 and R_3 generally are represented by R and it is very common to have lauric-lauric-oleic composition of fatty acid in the carbon chains (Scheme 2).

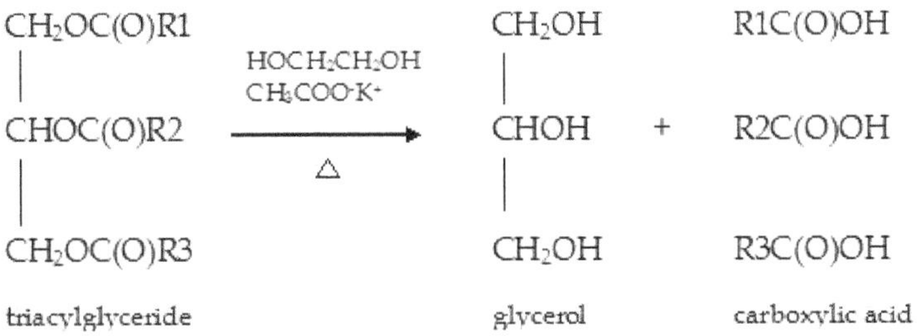

$$
\begin{array}{ccc}
\text{CH}_2\text{OC(O)R1} & & \text{CH}_2\text{OH} \qquad \text{R1C(O)OH} \\
| & \overset{\text{HOCH}_2\text{CH}_2\text{OH}}{\underset{\triangle}{\xrightarrow{\text{CH}_3\text{COO-K-}}}} & | \\
\text{CHOC(O)R2} & & \text{CHOH} \quad + \quad \text{R2C(O)OH} \\
| & & | \\
\text{CH}_2\text{OC(O)R3} & & \text{CH}_2\text{OH} \qquad \text{R3C(O)OH} \\
\text{triacylglyceride} & & \text{glycerol} \qquad\qquad \text{carboxylic acid}
\end{array}
$$

Scheme 1. Probable reaction mechanism between the RBD PKO and the hydroxyl-containing compound

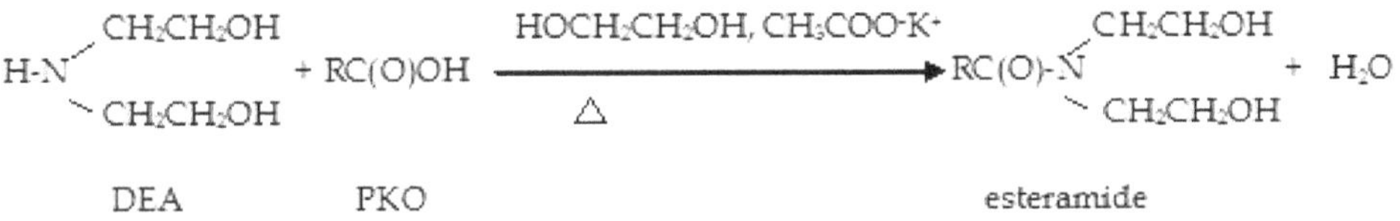

$$
\underset{\text{DEA}}{\text{H-N}\overset{\text{CH}_2\text{CH}_2\text{OH}}{\underset{\text{CH}_2\text{CH}_2\text{OH}}{}}} \; + \; \underset{\text{PKO}}{\text{RC(O)OH}} \; \xrightarrow[\triangle]{\text{HOCH}_2\text{CH}_2\text{OH, CH}_3\text{COO-K-}} \; \underset{\text{esteramide}}{\text{RC(O)-N}\overset{\text{CH}_2\text{CH}_2\text{OH}}{\underset{\text{CH}_2\text{CH}_2\text{OH}}{}}} \; + \; \text{H}_2\text{O}
$$

Scheme 2. Conversion of RBD PKO to the esteramide (RBD PKO-based polyol)

RBD PKO reacts with the polyhydroxyl compound in an alkaline medium (contributed by the potassium acetate). The alkalinity of the system ensured that the RBD PKO is fully reacted. The selection on polyhydroxyl compound being used is the critical part where it should offer highest hydroxyl value and functionality polyol possible to fully converting the RBD PKO into polyol (highest yield). Methods used in this study involved polycondensation and esterification where these are the only routes that offered low reaction temperature and short reaction time. It produced polyol (compound with functional group –OH) at high yield (almost 100%), low moisture content and no toxic vapor. The esteramide or PKO-based polyol is a monoester with OH terminal.

4. Properties of the PKO-based polyol

The derivatised RBD PKO-based polyol is a golden yellow liquid with a cloud point of 13°C. It has very low moisture content of 0.09% and low viscosity of 374 cps and specific gravity of 0.992 g/cm^3 at room temperature. Low water content and liquidy nature of the polyol are advantageous in formulating the polyurethane system especially when processing of end product is concerned. Less viscous polyol offers less viscous polyol resin which leads to system with good flowability. The viscosity increases as the degree of polycondensation and branching increases (Wood 1990). The physical properties of the PKO-based polyol are summarized in details in Table 4. It is important to note that raw RBD PKO solidified at room temperature with cloud point of about 23-24°C whilst the derivatized polyol solidified only at 13°C (cloud point). Polyol heating system is not required here as what is being used by other studies (Parthiban et al. 1999 and Ahmad et al. 1995).

Parameters	Result
State at 25°C	Liquid
Color	Golden yellow
Odor	Odorless
Density at 25°C, g/cm³	0.992
Solubility	Alcohol, Ketone, Ether, Alkane, Water
Cloud Point, °C	13
Viscosity at 25°C, cps	374
pH	9-10
Moisture content at 25°C, %	0.09

Table 4. Physical properties of the derivatised RBD PKO-based polyol.

4.1. Chemical analysis

a.   Fourier Transform Infrared Spectroscopy (FTIR)

The RBD PKO, a chain of fatty acid with carboxylic acid group displays intense C=O stretching bands of acids absorb at 1711 cm^{-1} as shown in Fig. 3 (a). The C-H stretches at 2932 and 2855 cm^{-1}. Two bands arising from C-O stretching and O-H bending appear in the spectra of RBD PKO near 1320-1210 and 1440-1395 cm^{-1} respectively. Both of these bands involve some interactions between C-O stretching and in-plane C-O-H bending. The C-O-H bending band near 1440-1395 cm^{-1} is of moderate intensity and occurs in the same region as the CH$_2$ scissoring vibration of the CH$_2$ group adjacent to the carbonyl (Silverstein et al. 1991).

The FTIR spectrum of the derivatized RBD PKO was obtained from samples taken at 175-180°C (Fig. 3(b)) during the esterification process. The spectrum was evaluated at peak 3351 cm^{-1} (designated as peak A) and 1622 cm^{-1} (designated as peak B). Peak A and B, which are the hydroxyl (-OH) and carbamate (O=C=N-) peaks respectively (assigned by IR Mentor Pro Classes, Sadtler Division Bio-Rad Laboratories 1990 and Silverstein et al. 1991). These peaks do not appear in the spectra of the raw RBD PKO (Fig. 3(a)). A vague trace of the hydroxyl peak was observed when PKO is mixed with the hydroxyl compound. Further increase in the reaction temperature and reaction time changed the percentage of transmittance for both peaks A and B significantly. It also indicated a formation of ester cleavage at 1710 cm^{-1}. The sharp absorption bands in the region of 1750-1700 cm^{-1} are characteristic of carbonyl group of ester (C=O) stretching vibrations (Silverstein et al. 1991).

Transmittance ratio of both peaks, the OH and the carbamate peaks (% transmittance of peak A divided by the % transmittance of peak B) was plotted as in Fig. 4. It was used to identify the progress of the derivatization process (Chian and Gan 1998). Fig. 4 also showed that the hydroxyl value (OHV) reached to a constant at 350-370 mg KOH/g sample at intervals of 175-180°C for 15-30 minutes of reaction time. The FTIR spectrum and hydroxyl value (OHV) curves both demonstrated that 175-180°C at 15-30 minutes as optimum temperature and reaction time respectively. Both methods are advantageous in the identification of optimum processing parameters assuming that Beer's Law is applicable here. However, OHV determination method is slow and time-consuming. Therefore, FTIR method is more preferable in determining the completion of reaction for the RBD PKO-based polyol (Chian and Gan 1998).

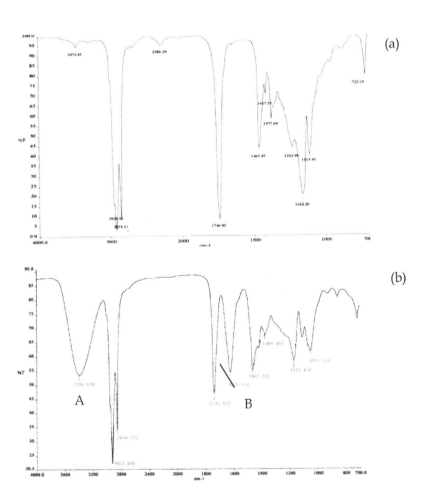

Figure 3. FTIR spectra of (a) the raw RBD PKO and (b) the palm-based esteramide

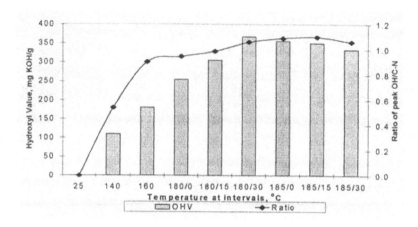

Note:
25 refers to derivatized RBD PKO at ambient temperature, 25°C
140 refers to derivatized RBD PKO at 140°C
160 refers to derivatized RBD PKO at 160°C
180/0 refers to derivatized RBD PKO at starting of 180°C
180/15 refers to derivatized RBD PKO at 180°C for 15 minutes
180/30 refers to derivatized RBD PKO at 180°C for 30 minutes
185/0 refers to derivatized RBD PKO at starting of 180°C
185/15 refers to derivatized RBD PKO at 185°C for 15minutes
185/30 refers to derivatized RBD PKO at 185°C for 30 minutes

Figure 4. Curve of ratio of OH peak to the C-N peak and the OHV curve of the blend at intervals

b.   Thin Layer Chromatography

The thin layer chromatography (TLC) test on the desired products obtained at intervals of
reaction time at 175-180°C (0, 15 and 30 minutes) indicated a clear qualitative separation.
These separations were compared to TLC carried out on individual ingredients: The
RBD PKO, diethanolamine (DEA), the catalyst-potassium acetate in monoethylene glycol
and standard lauric acid (Athawale et al. 2000). There were three separation peaks,
identify as the PKO, DEA and small trace of the catalyst up to 175-180°C at 0 minute. At
175-180°C for 15 minutes, only two separation peaks were observed and finally at 175-
180°C for 30 minutes, only one separation peak was observed (Fig. 5). The result is parallel
to the gas chromatography (GC) peaks of the final product, the RBD PKO-based polyol
(Fig. 6)).

c.   Gas Chromatography-Mass Spectrometry (GC-MS)

The samples collected at intervals ranging from 165-170°C, 170-175°C, 175-180°C and
180-185°C were also evaluated for its purity using gas chromatography, GC coupled with
mass spectrometry, GC-MS. Fig. 6 is the GC of the RBD PKO-based polyol reacted at 175-
180°C for 15-30 minutes. The signal at retention time of 31.92 min is the desired product,
the RBD PKO-based polyol (98.24%) while signals at retention time of 13.37 (0.08%), 16.36
(0.92%) and 27.91 (0.27%) representing small percentage traces of MEG, glycerol (by-

product of esterification) and DEA (C:\ DATABASE\WILEY275.L). Others (0.49%) are traces of oligomeric polyester components from C_{14} and C_{18} chains. The GC-MS scan of the RBD PKO-based polyol showed an estimated molecular weight of 477. Molecular weight obtained at 165-170 and 170-175°C of reaction temperature was 296 and 355 respectively. Thus, molecular weight obtained at 175-180°C is considered to be the most desirable molecular weight for this study. The functionality of the RBD PKO-based polyol derived from this molecular weight and the determined hydroxyl value (OHV of 350 to 370 mg KOH/g) is 2.98 to 3.15 calculated using the mathematical equation in equation 3.

$$\text{Functionality} = M_w \times OHV / 56100 \qquad (3)$$

Note: M_w is the estimated molecular weight of the RBD PKO-based polyol obtained from GC-MS which is 477 OHV is the hydroxyl value of the RBD PKO-based polyol obtained using ASTM D4274-88 method, which is about 350-370 mg KOH/g sample

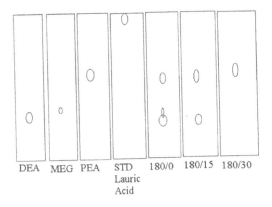

DEA MEG PEA STD   180/0   180/15 180/30
                Lauric
                Acid

Note:
PEA   RBD PKO-based polyol
DEA   diethanolamine
MEG   monoethylene glycol
STD   standard lauric acid
180/0   derivatised RBD PKO at starting of 180°C
180/15   derivatised RBD PKO at 180°C for 15 minutes
180/30   derivatised RBD PKO at of 180°C for 30 minutes

Figure 5. The thin layer chromatography of the ingredients

This range of functionality is suitable for rigid foam application (Wood 1990).

Both FTIR (IR Mentor Pro 1990) and GC-MS approaches (Wiley MS-database) could be used to estimate the most probable molecular structure of the RBD PKO-based polyol at 175-180°C/30 minutes (optimum temperature and reaction time) as 2-hydroxy-undecanoamide as in Scheme 2 (library search on Wiley MS-database giving 98% quality match). There is no intention of purification of the synthesized RBD PKO-based polyol as all these hydroxyl-containing compounds would react with crude MDI.

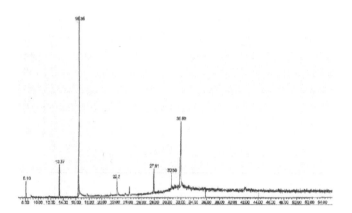

Figure 6. GC chromatogram of the RBD PKO-based polyol obtained at 175-180ºC for 30 minutes.

4.2. Thermal testing

The thermogram of the resulted RBD PKO-based polyol is as shown in Figure 7. Thermally, it is stable up to 167.6ºC and undergoes two stages decomposition at 167.6 to 406.3ºC with total weight loss of 99.41%. The initial 3.34% weight loss is contributed to the moisture content and other volatile impurities in the RBD PKO-based polyol (Oertel 1993). The initial decomposition is contributed by the degradation of RBD PKO-based polyol and traces of glycerol supported by the DTA curve which representing the softening temperature at 385ºC. Charred residue was obtained after testing.

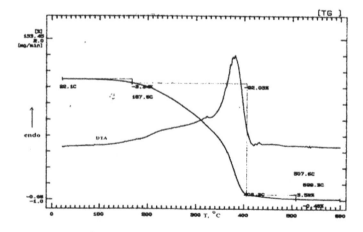

Figure 7. TGA thermogram of the RBD PKO-based polyol obtained at 175-180ºC for 30 minutes

5. Properties of the PKO-based polyurethane foam

5.1. Physical properties

The PKO-based polyurethane foam (PUF) produced is a light yellow solid with skin thickness of about 1.5 mm. It is a stiff/rigid but brittle solid at 43-44kg/m^3 molded density and core density of 38-39 kg/m^3 with average void size of 0.10-0.15 mm (Fig. 8).

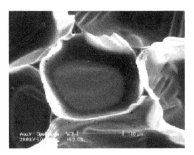

Figure 8. Scanning electron micrograph of the PUF at 250× magnification

5.2. FTIR analysis

The PUF is analysed by FTIR spectroscopy which showed the absence of the free OH groups and indicates a complete conversion of both –OH groups of the PEA to the urethane moiety (NH-C(O)-O). Typical FTIR spectrum of the PU is as shown in Fig. 9. The characteristic –NH stretching vibration of the –NH$_2$- (amide) is located at 3405 cm^{-1}, overlapping with the OH peak as a broad band. Bands at 2932 and 2894 cm^{-1} are the synchronous reflection of asymmetric and symmetric of CH$_2$ bridges, from the linkage of the urethane with the PEA. Bands at 1650 cm^{-1} is the overlapping of –N=C=O (urethane) and ester linkage of the PEA. Obviously, bands 1550, 1650 and 3350 cm^{-1} indicate complete conversion to urethane moiety (Silverstein et al. 1991).

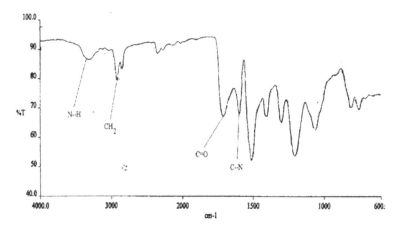

Figure 9. FTIR spectrum of the RBD PKO polyurethane foam

5.3. Thermal properties

The thermal instability of a PU may be defined as the ease by which heat produces changes in the chemical structure of the polymer network. These may involve simple bond-rupturing dissociation or reaction reversals and provide more volatile components, or they may result in extensive pyrolysis and fragmentation of the polymer. This characteristic provides a measure of fire hazard in that a more thermally stable polymer is less likely to ignite and contribute to a conflagration than a less stable one (Burgess, Jr. & Hilado 1973). Thermodynamic parameters such as decomposition temperatures, percentage of weight loss, melting temperature, T_m and glass transition temperature, T_g were determined by thermal analyses of the PU.

TGA thermogram of the PU is as shown in Fig. 10. Presence of three degradation stages implying the presence of three thermal degradation temperatures. It was thermally stable at 191.9ºC, a common stability temperature for PU (Hepburn 1991). The initial weight loss of about 41.24% commences at 191.9 to 396.9ºC. T_{max} from the DTA curve occurred at 275ºC attributed by carbon dioxide trapped in the sample. Degradation started at 396.3 to 498.4ºC, which was initially a fast process. The total weight loss up to 500ºC is 74%. This second stage of degradation rationalized the urethane linkage reported by Hepburn (1991).

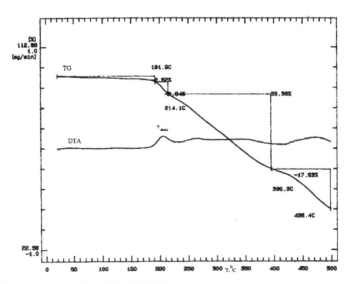

Figure 10. TGA thermogram of the RBD PKO PU foam

There is no indication of melting and crystallization temperatures curves in the DSC thermogram of the PU foam. Therefore, this polymer could be identified as an amorphous polymer (Badri et al. 2000). The glass transition temperature is 39.74ºC, a regular glass transition temperature for thermosetting polymers, with heat capacity of 33.0 J/g. Since the heat of evaporation of water is 2300J/g, moisture content of the PU was detected to be 1.43%.

However, the initial thermal conductivity of PU foam is found to be 0.0244 W/m-K. This is expected since it is a water-blown PU system where water has poor properties i.e. boiling point and k-factor compared to its industrial counterpart, chlorofluorocarbon, CFC (Crawford & Escarsega 2000). Low thermal conductivity is normally contributed by the low boiling point of the blowing agent such as CFC and finer cell structure of the foam (Hardings 1965, Frisch 1985, Hepburn 1991). However, another fact to be remembered is that water is a chemical blowing agent compared to CFC, a physical blowing agent. Water is capable of interfering in the polymerization of polyurethane by producing by-product such as urea and large amount of carbon dioxide when being used I larger quantity. Functionality of polyols also plays an important role in producing good insulated material (Wood 1990, Hass and Uhlig 1985).

5.4. Mechanical properties

The mechanical properties of the foam produced from the derivatized RBD PKO show comparable results (Table 5) to the British Standard requirement (practiced by industry such as building construction industry). It is expected for water-blown PU foam to have lower compressive stress at 5% strain and the compression due to irregular formation of cellular structure. This decreased the strength upon higher percentage of strain. Table 5 showed the summary of the mechanical properties of the PU foam.

The dimensional stability which is described in percentage of volume change indicated changes of -0.090% and 0.012% at -15±2ºC and 70±2ºC at 95±5% relative humidity for 24 hours respectively. A very minimum shrinkage and expansion problem was observed on the foam prepared from this palm oil-based polyurethane polyol in a water-borne system. Identical resin formulation was used using petroleum-based polyol to substitute the palm-based polyol. Major shrinkage and expansion problems were observed. Shrinkage and expansion problems are normally used as indicators of how good the foam is as an insulator. The mechanical properties could be enhanced by using low or high pressure dispersing machines (Oertel 1993). Better mechanical properties could also be achieved by introducing filler in the PU system (Rozman et al. 2001[a], 2001[b], 2000, 1998).

5.5. Rheological and kinetic properties

The PU system is polymerized kinetically using tetramethylhexadiamine, TMHDA as a gel/blow catalyst and pentamethyldiethylenetetramine, PMDETA as a blow catalyst. The addition of both catalysts is very minimum (0.05-0.10 pbw) in achieving an optimum kinetic reaction time (Tamano et al. 1996) especially when reactive RBD PKO-based polyol (Scheme 2) is used in the formulation. The cream time, gelling/fiber time, tack-free time and rise time (Appendix B) were 23, 71, 105 and 156 seconds respectively at 20ºC. The PUF is demolded after 10 minutes of mixing with skin thickness of about 1.5 mm. It has a flow index of 1.050 cm/g, a moderate flowability PU system (Colvin 1995). This is assumed to be helpful in reducing the consumption of raw materials, especially the RBD PKO-based polyol.

5.6. Resistance to environmental stress

The chemical resistance of the PU with normal closed-cell structures of rigid urethane foam prepared from the crude MDI and RBD PKO-based polyol is carried out to investigate the limitation of the interactions with surroundings to the surface layer in order to produce a chemically and physically stable material. Effects produced by chemical agents depend both on the chemicals and on the permeability of cell membranes. Solubility of the chemical in the foam affects both permeability and swelling. Results obtained are not representative of other temperatures, concentrations or exposure times.

Parameter	Method	Standard	Results
*Apparent molded density, kg/m^3	BS 4370:Part 1:1988 (Method 2)	Min 38	43.6 ±0.85
*Apparent density (core), kg/m^3	BS 4370:Part 1:1988 (Method 2)	Min 35	38.9 ±0.53
*Compressive strength to foam rise at 10% deflection, kPa	BS 4370:Part 1:1988 (Method 3)	Min 180	185.7 ±8.22
*Compressive stress at 5% strain, kPa	BS 4370:Part 2: 1993 (Method 6)	Min 140	105.4 ±2.41
Compressive modulus, N/m^2	BS 4370: Part 1: 1988 (Appendix A)	Not available	8.52 ±0.46
*Dimensional stability,%	BS 4370: Part 1: 1988 (Method 5B)		
	At -15 ±2°C for 24h	Maximum 1.0	Length: -0.151 ±0.03 Width: -0.433 ±0.03 Thickness: 1.373 ±0.06
	At 70 ±2°C, 95 ±5% r.h. for 24h	Maximum 3.0	Length: 0.359 ±0.25 Width: 0.017 ±0.04 Thickness: 1.654 ±0.09
*Apparent water absorption,%	BS 6586: Part 1: 1993 (Annex D)	Maximum 6.5	2.25 ±0.89
Shore A Hardness	ASTM D 2240	Not Available	29.0 ±1.4

Note: * Physical property requirements following BS6586: Part 1: 1993 industrial standard.

Table 5. The mechanical properties of the PU foam synthesized from the RBD PKO-based polyol.

Fig. 11 illustrates the compressive strength at 10% deflection and 5% strain as well as its compression modulus upon exposure to stress. All resistivity test medium being used result in a major increment in the strength at 10% deflection. Readings of above 0.20MPa (compared to the control foam) with maximum compressive strength are observed in benzene at about 0.34MPa, followed by PUF at ambient temperature (0.30MPa), freeze-thaw condition (0.26MPa), 10% NaOH (0.25MPa), saltwater (0.20MPa) and finally 10% HNO₃ (0.19MPa). The same trend is observed in compressive strength at 5% strain where the maximum value is encountered at freeze-thaw condition followed by at ambient temperature, 10% NaOH and finally benzene. The compression modulus reaches as high as 11.0MPa and others are in the range of 8.0 to 9.0MPa.

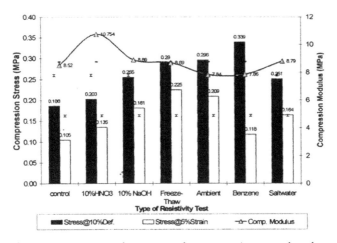

Figure 11. Effect of various environmental stresses on the compressive strength and compression modulus of the RBD PKO-based PU foam

Practically, the absorption of chemicals into the foam results in swelling of the cell faces, which apparently increases the compressive strength. Weathering conditions (ambient and freeze-thaw) however are very much dependence on the diffusion rate of carbon dioxide being replaced by the air which causes expansion of the foam and increases the compressive strength (Wood 1990). The foams are found to be unaffected by the test medium basically due to the mixture of organic components (RBD PKO-based polyol and MDI). Rigid PU foam is stable in the present of most solvents such as found in binders and sealers (Oertel 1993).

Physically, the foam becomes spongy with the formation of waxy material on the surface of the foam, as a result of prolonged exposure to benzene as an aromatic hydrocarbon. It is important to note that ester-based polyurethanes are easily attacked by hot aqueous alkali or moderately concentrated mineral acids, swollen by aromatic hydrocarbons and decomposed by prolonged contact with water, diluted acids and moist heat (causes swelling and slow hydrolysis) (Roff et al 1971).

The compression modulus of the PUF ranges from 7.8 to 10.8MPa. the compression modulus for the control PUF is at 8.5MPa which is lower compared to the modulus in 10%HNO₃,

10%NaOH, under freeze-thaw condition, and in saltwater but higher if compared to the modulus of the rest of the resistivity test.

Rigid PU prepared has high resistivity to the action of most organic solvents and are seriously degraded only by strong acid, oxidizing agent and corrosive chemicals. Only stronger polar solvents, which significantly swelled the polymer, led to shrinkage of the foam structure. Evaporation of the solvent normally returns the polymer to its original state (Oertel 1993).

6. Conclusion

Several advantages are foreseen from this study. Some important advantages are being identified through this method of polyol production. Firstly, it is attractive and economical. RBD PKO-based polyol is a naturally formed macromolecules found in Malaysia. It is extremely plentiful, easy to process and refine, capable of being cultivated with minimum capital investment and suitable for conversion to quality polyols using an inexpensive reaction process. Secondly is the simplicity of the process, which requires only a few reactors for producing the polyol as well as formulating the resin. Commercially, the process acquires only a few personnel to produce consistently good quality polyols. Thirdly, compare to the manufacturing of the petrochemical-based polyols, the process is relatively safe, where it involves the usage of hazardous chemicals. Generally, it is non-toxic and of low volatility.

Two major environmental advantages can be realized. Firstly, the source of oil is truly renewable, where it does not lead to permanent depletion of resources which has a limited global availability. Secondly, the amount of energy required to convert the natural oils to polyol is considerably less than using the conventional process. The foam made from this RBD PKO-based polyol is low in density, light in color, high in strength but low in water sorption. The produced RBD PKO-based polyurethane foam in this study also has other advantages as tabulated in Table 6.

Property	Rating	Consequence	Benefit
Thermal Insulation	Highest	Thinnest Section	Space
Rigidity	High	Added Strength	Structural
Adhesion	High	No glue-line	Manufacturing
Dimensional Stability	High	Non-sag, non-heave	Maintenance
Density	Low	Lightweight	Handling
Water Vapor Transmission	Low	Less Condensation	Construction

Table 6. The advantages of producing RBD PKO-based polyurethane.

The PUF meets the British Standard requirements in any medium of the tested environmental stress test. This ester-type polyurethanes are easily attacked by prolonged contact with water, diluted acids and moist heat (causes swelling and slow hydrolysis) and swollen by aromatic hydrocarbons. These rigid PUs either the PUF, are resistant to the action of most organic solvents and are seriously degraded only by strong acids, oxidizing agents and corrosive chemicals. Only polar solvents, which significantly swell the polymer, lead to shrinkage of the foam structure. Evaporation of the solvent normally returns the polymer to its original state.

In terms of application, these composites are most suitable in structures where stiffness and dimensional stability are of prime importance but is only a secondary choice to areas where structural strength is more vital than the component rigidity.

Author details

Khairiah Haji Badri
Polymer Research Center, Faculty of Science and Technology,
Universiti Kebangsaan Malaysia, Selangor, Malaysia

Acknowledgement

These works on the production of the RBD PKO-based polyol and other ranges of polyurethane polyols have been at present being produced at larger scale and ready to depart January 2012. This is being brought into realization with the support of Universiti Kebangsaan Malaysia under its entities School of Chemical Sciences and Food Technology, Polymer Research Center and Faculty of Science and Technology (UKM-OUP-FST-2012) for all the facilities provided. Thank you to Ministry of Higher Education, Ministry of Science, Technology and Innovation (previously known as Ministry of Science, Technology and Environment) and Yayasan Felda for the financial supports. Major contributions definitely came from graduates and colleagues of Universiti Kebangsaan Malaysia. For special individuals who initiated this project, Zulkefly Othman and in memory Haji Badri Haji Zakaria, my greatest thanks to both of you.

7. References

Ahmad, S., Siwayanan,P. & Wiese, D. 1995. Porim and INTERMED Sdn.Bhd. Malaysian Patent Application Number. PI9502302. Filling Date: 7 August, 1995.

Alfani, R., Iannace, S.&Nicolois, L. 1998. Synthesis and Characterization of Starch Based Polyurethane Foams. *J. Appl. Polym.* Sci. 68 (5) : 739-745

Apukhtina, N.P. 1973. Methods for Increasing the Thermal Stability of Polyurethanes: Soviet Urethane Technology, Ed. Schiller, A.M. pp. 198-210. Connecticut: Technomic Publishing Co., Inc.

Arnold, J.M. 1983. *Vegetable Oil Extended Polyurethane System.* US 4375521

Athawale, V.D., Rathi, S.C. & Bhabe, M.D. 2000. Novel Method For Separating Fatty Ester From Partial Glycerides in Biocatalytic Transesterification Of Oils, *Separation and Purification Technology,* 18:3:209-215.

Austin, P.E., Derderian, E.J.& Kayser, R.A. 2000. Hydrosilation in High Boiling Natural Vegetable Oils. US 6071977.

Norzali N.R.A., Badri, K.H. & Nawawi, M.Z. 2011[a]. Loading Effect of Aluminum Hydroxide onto the Mechanical, Thermal Conductivity, Acoustical and Burning Properties of the Palm-based Polyurethane Composites, Sains Malaysiana 40(7): 737-742

Norzali N.R.A., Badri, K.H. & Nawawi, M.Z. 2011[b]. The Effect Of Aluminium Hydroxide Loading On The Burning Property of The Palm-Based Polyurethane Hybrid Composite, Sains Malaysiana 40(4):385–390.

Liow, C.H., Badri, K.H. & Ahmad, S.H. 2011. Mechanical and Thermal Properties of Palm-Based Polyurethane Composites Filled With Fe3O4, PANI and PANI/F e3O4, Sains Malaysiana 40(4): 379–384.

Wong, C. S. & Badri, K.H. 2010. Sifat Terma Dan Kerintangan Api Poliuretana Berasaskan Minyak Isirung Sawit Dan Minyak Kacang Soya, Sains Malaysiana, 39 (5): 775-784.

Badri, K.H. & Redhwan, A. M. 2010. The effect of phosphite loading on the mechanical, thermal and fire properties of palm-based polyurethane, Sains Malaysiana, 39 (5): 769-774.

Mat Amin, K. A., Badri, K.H. & Othman, Z. 2007. Oil Palm-Based Hybrid Biocomposites with Kaolinite. Journal of Applied Polymer Science 105:2488-2496.

Badri, K.H., Ujar, A. H., Othman, Z. & Sahaldin, F. H. 2006[a]. Shear Strength of Wood-to-Wood Adhesive Based on Palm Kernel Oil, Journal of Applied Polymer Sciences, 100(3): 1750-1759

Badri, K.H., Mat Amin, K. A., Khalid, N. K., Othman, Z. & Abdul Manaf, K. 2006[b]. Effect Of Filler-To-Matrix Ratio On The Mechanical Strength Of Palm-Based Biocomposite Board, Polymer International, 55: 190-195

Badri, K.H. & Mat Amin, K. A. 2006[c]. Oil Palm-Based Biocomposites. Journal of Oil Palm Research, (Special Issue-April 2006):103-113

Badri, K.H., Othman, Z. & Mohd Razali, I. 2005. Mechanical properties of polyurethane composites from oil palm resources. Iranian Polymer Journal, 14 (5): 987-993

Badri, K.H., Othman, Z. & Ahmad, S.H. 2004[a]. Rigid Polyurethane Foams From Oil Palm Resources, Journal Of Materials Science. 39(16-17):5541-5542

Badri, K.H., Shahaldin, F. H. & Othman, Z. 2004[b]. Indigenous Coating Material From Palm Oil-Based Polyamide. J. Mater. Sci. Letters,39 (13):4331-4333.

Khairiah Haji Badri. 2002. Preparation and Charaterization of Polyurethane Foam from RBD Palm Kernel Oil-Based Polyurethane Polyol and Oil Palm Empty Fruit Bunch Fiber As Filler. *Proceeding of National Science Fellowship (NSF) Workshop*, pp 114-120.

Badri, K.H., Ahmad, S.H & Zakaria , S. 2000[a]. Development of Zero ODP Rigid Polyurethane Foam From RBD Palm Kernel Oil: *J. Mater. Sci. Letters*, 19: 1355-1356.

Badri, K.H, Ahmad, S.H. & Zakaria, S. 2000[b]. Thermal, Crystallinity and Morphological Studies on the Filled RBD Palm Kernel Oil Polyurethane Foam: *Nuclear Science Journal of Malaysia*, 18 (2): 57-62.

Badri, K.H., Ahmad, S.H. & Zakaria, S. 2001[a]. Production of a High-Functionality RBD Palm Kernel Oil-Based Polyester Polyol. *J. Appl. Polym. Sci.*, 81 (2): 384-389.

Benli,S., Yilmazer, U., Pekel, F. & Ozkar, S. 1998. Effect of Fillers on Thermal and Mechanical Properties of Polyurethane Elastomer, *J. Appl. Polym. Sci.* 68: 1057-1065.

Berlin, A.A. & Zhitinkina, A.K. 1982. Foam Based on Reactive Oligomers, Polyurethane Foams, pp. 51-111. London: Howard Publishing Inc.

Bhunia, H.P., Jana, R.N., Basak A., Lenka, S. & Nando, G.B. 1998. Synthesis of Polyurethane From Cashew Nut Shell Liquid (CNSL), A Renewable Resource. *J. Appl. Polym. Sci.* 36 (3): 391-400.

Burgess, Jr., P.E & Hilado, C.J. 1973. Thermal Decomposition and Flammability of Foams: *Plastic Foams* Part II. Ed. Frisch, K.C & Saunders, J.H. pp. 855-871. New York: Marcel Dekker, Inc.

C:\Database\wiley275.1.1999. Library Search of Acquisition Method. University of Malaya.

Chian, K.S. & Gan, L.H. 1998. Development of a Rigid Polyurethane Foam From Palm Oil. *J. Appl. Polym. Sci.*, 68 (3): 509-515

Chittolini, C. 1999. *Polyurethane Foam-Mixing Isocyanate Component and Polyol Component Including Pentane and Dialkanolamine Derived from Vegetable Oil or Fat to Make Polyurethane Foam.* US 5859078

Colvin, B.G. 1995. Low Cost Polyols From Natural Oils, *U'tech Asia '95.* 36: 1-10

Crawford, D.M. & Escarsega, J.A. 2000. Dynamic Mechanical Analysis of Novel Polyurethane Coating for Military Applications. *Thermochimica Acta*, 357-358: 161-168

Cunningham, R.L., Gordon, S.H., Felker, F.C. & Eskins, K. 1997. Jet-Cooked Starch Oil Composite in Polyurethane Foams. *J. Appl. Polym.Sci.* 64 (7): 1355-1361

Frisch, K.C. 1985. Fundamental Chemistry and Catalysis of Polyurethanes, *Polyurethane Technology*, Ed. Bruins, P.F. pp. 12-17. New York: Interscience Publishers.

Harding, R.H. 1965. Effect of Cell Geometry On PU Foam Performance, *J. Cell. Plastics*, 1: 224

Hass, P.F. & Uhlig, K. 1985. Additive and Auxiliary Materials, *Polyurethane Handbook,* 2nd ed. Ed. Oertel, G. pp. 98. New York: Hanser Publisher.

Hepburn, C. 1991. Polyurethane Elastomers. 2nd ed., pp. 441: Great Britain: Elsevier Science Publishers Ltd.

IR Mentor Pro Classes. 1990. Bio-Rad Laboratories, Sadtler Division, PEIM: 6

Javni,I., Petrovic, Z.S., Guo, A. & Fuller, R. 1999. Thermal Stability of Polyurethane-Based on Vegetable Oils. Annu. Tech. Conf.-Soc. Plast. Eng. 3: 3801-3805

Jedlinski, Z.J. 1992. Polyester: Handbook of Polymer Synthesis Part A, Ed. Kricheldorf, H.R. pp. 645-648. London: McGraw-Hill.

Loudon, G.M. 1988. Chemistry of Carboxylic Acids, *Organic Chemistry.* 2nd ed.pp. 816-817. London: Mc Graw Hill.

Mohapatra, D.K., Das, D., Nayak, P.L. & Lenka, S. 1998. Polymers From Renewable Resources. XX. Synthesis, Structure And Thermal Properties Of Semi- Intherpenetrating Polymer Networks Based On Cardanol-Formaldehyde Substituted Aromatic Compounds Copolymerized Resins And Castor Oil Polyurethanes. *J. Appl. Polym. Sci.* 70 (5): 837-842

Nayak, P., Mishra, D.K., Parida, D., Sahoo, K.C., Nanda, M. Lenka, S. & Nayak, P.L. 1997. Polymers From Renewable Resources. IX. Interpenetrating Polymer Networks Based On Castor Oil Polyurethane Poly(hydroxylethylmetacrylate): Synthesis, Chemical, Thermal and Mechanical Properties. *J. Appl. Polym. Sci.* 63 (5): 671-679

Oertel, G. 1993. *Polyurethane Handbook: Chemistry-Raw Material-Processing- Application-Properties. Cincinnati:* Hanser Gardner Publications, Inc.

Parthiban, S., Ooi, T.L., Kassim Shaari, N.Z., Ahmad, S., Wiese, D. & Chua, M.C. 1999. Polyurethane From Palm-Based Polyols, *Palm Oil Technical Bulletin,* September-October 1999: 4-6

Pourjavadi, A., Rezai, N. & Zohuriaan-M, M.J. 198. A Renewable Polyurethane: Synthesis and Characterization of the Interpenetrating Networks (IPNs) From Cardanol Oil. *J. Appl. Polym. Sci.,* 68: 173-183

Roff, W.J., Scott, J.R. & Pacitti, J. 1971. Fibres, *Films, Plastics and Rubbers: A Handbook of Common Polymers.* Pp. 446-457. New York: Butterworth & Co. (Publishers) Ltd.

Rozman, H.D. Tay, G.S., Kumar, R.N., Abusamah, A., Ismail. H. & Mohd Ishak, Z.A. 2001a. Polypropylene-Oil Palm Empty Fruit Bunch- Glass Fibre Hybrid Composites: A Preliminary Study on the Flexural and Tensile Properties. *European Polymer Journal.*37 (6): 1283-1291.

Rozman, H.D., Tay, G.S., Abubakar, A. & Kumar, R.N. 2001b. Tensile Properties of Oil Palm Empty Fruit Bunch- Polyurethane Composites. European Polym. Journ., 37: 1759-1765

Rozman, H.D., Lai, C.Y., Ismail, H. & Mohd Ishak, Z.A. 2000. The Effect Of Coupling Agents On the Mechanical And Physical Properties Of Oil Palm Empty Fruit Bunch-Polypropylene Composites. *Polym. Int.* 49 (11): 1273-1278

Rozman, H.D., Kon. B.K., Abusamah, A., Kumar, R.N. & Mohd Ishak, Z.A. 1998. Rubberwood-High Density Polyethylene Composites: Effect of Filler Size and Coupling Agents on Mechanical Properties. *J. Appl. Polym. Sci.* 69: 1993-2004

Silverstein, R.M., Bassler, G.C. & Morril, T.C. 1991. *Spectrometric Identification of Organic Compounds,* 5th ed., New York: John Wiley & Sons, Inc

Tamano, Y., Yoshimura, H., Ishida, M., Okuzono, S. & Lowe, D.W. 1996. The Characteristics and Role of tertiary Amine catalysts For Polyurethane Foams: Review of Tertiary Amine Catalysts "TEDA & TOYOCAT", *Conference Paper of UTECH '96*

Wood, G. 1990. The Chemistry and Materials of PU Manufacture, *The ICI Polyurethane Book,* 2nd *ed.* Ed. Genge, R. & Sparrow, D. pp. 41-42. New York: John Wiley & Sons.

Permissions

The contributors of this book come from diverse backgrounds, making this book a truly international effort. This book will bring forth new frontiers with its revolutionizing research information and detailed analysis of the nascent developments around the world.

We would like to thank Eram Sharmin and Fahmina Zafar, Ph.D., for lending their expertise to make the book truly unique. They have played a crucial role in the development of this book. Without their invaluable contribution this book wouldn't have been possible. They have made vital efforts to compile up to date information on the varied aspects of this subject to make this book a valuable addition to the collection of many professionals and students.

This book was conceptualized with the vision of imparting up-to-date information and advanced data in this field. To ensure the same, a matchless editorial board was set up. Every individual on the board went through rigorous rounds of assessment to prove their worth. After which they invested a large part of their time researching and compiling the most relevant data for our readers. Conferences and sessions were held from time to time between the editorial board and the contributing authors to present the data in the most comprehensible form. The editorial team has worked tirelessly to provide valuable and valid information to help people across the globe.

Every chapter published in this book has been scrutinized by our experts. Their significance has been extensively debated. The topics covered herein carry significant findings which will fuel the growth of the discipline. They may even be implemented as practical applications or may be referred to as a beginning point for another development. Chapters in this book were first published by InTech; hereby published with permission under the Creative Commons Attribution License or equivalent.

The editorial board has been involved in producing this book since its inception. They have spent rigorous hours researching and exploring the diverse topics which have resulted in the successful publishing of this book. They have passed on their knowledge of decades through this book. To expedite this challenging task, the publisher supported the team at every step. A small team of assistant editors was also appointed to further simplify the editing procedure and attain best results for the readers.

Our editorial team has been hand-picked from every corner of the world. Their multi-ethnicity adds dynamic inputs to the discussions which result in innovative outcomes. These outcomes are then further discussed with the researchers and contributors who give their valuable feedback and opinion regarding the same. The feedback is then collaborated with the researches and they are edited in a comprehensive manner to aid the understanding of the subject.

Apart from the editorial board, the designing team has also invested a significant amount of their time in understanding the subject and creating the most relevant covers. They scrutinized every image to scout for the most suitable representation of the subject and create an appropriate cover for the book.

The publishing team has been involved in this book since its early stages. They were actively engaged in every process, be it collecting the data, connecting with the contributors or procuring relevant information. The team has been an ardent support to the editorial, designing and production team. Their endless efforts to recruit the best for this project, has resulted in the accomplishment of this book. They are a veteran in the field of academics and their pool of knowledge is as vast as their experience in printing. Their expertise and guidance has proved useful at every step. Their uncompromising quality standards have made this book an exceptional effort. Their encouragement from time to time has been an inspiration for everyone.

The publisher and the editorial board hope that this book will prove to be a valuable piece of knowledge for researchers, students, practitioners and scholars across the globe.

List of Contributors

Eram Sharmin and Fahmina Zafar
Materials Research Laboratory, Department of Chemistry, Jamia Millia Islamia (A Central University), New Delhi, India

Ruslan Davletbaev, Ilsiya Davletbaeva and Olesya Gumerova
Kazan National Research Technological University, Russia

Mohammed Ahmed Issam and Hamidi Mohamed Rashidah
University Sains Malaysia, Malaysia

Nataly Kozak, Anastasyia Hubina and Eugenia Lobko
Institute of Macromolecular Chemistry, National Academy of Sciences of Ukraine, Ukraine

Ahmadreza Gharehbagh
Iran Polyurethane Mfg. Co. NO.30, Tehran, Iran

Zahed Ahmadi
Color and Polymer Research Center, Amirkabir University of Technology, Tehran, Iran

Suzana M. Cakić and Olivera Z. Ristić
University of Niš, Faculty of Technology, Leskovac, Serbia

Ivan S. Ristić
University of Novi Sad, Faculty of Technology, Novi Sad, Serbia

Khairiah Haji Badri
Polymer Research Center, Faculty of Science and Technology, Universiti Kebangsaan Malaysia, Selangor, Malaysia

Printed in the USA
CPSIA information can be obtained
at www.ICGtesting.com
JSHW011355221024
72173JS00003B/294